BEEKEEPING IN THE UNITED STATES

UNITED STATES
DEPARTMENT OF
AGRICULTURE

AGRICULTURE
HANDBOOK
NUMBER 335

PREPARED BY
SCIENCE AND
EDUCATION
ADMINISTRATION

Books for Business
New York-Hong Kong

Beekeeping in the United States

by
United States Department
of Agriculture

ISBN: 0-89499-163-9

Reprinted from the 1980 edition

Books for Business
New York - Hong Kong
http://www.BusinessBooksInternational.com

CONTENTS

INTRODUCTION

By E. C. Martin [1]

Beekeeping is an ancient art that has fascinated its devotees since earliest times. Honey robbed from wild colonies in trees or caves was early people's main source of sweet food. Dominance of honey as the major sweetener continued until cane and beet sugar became generally available in comparatively recent times. Honey with its unique flavors and aromas and natural origin still has wide appeal. World production was estimated at 1,446 million pounds in 1976 and more than 1,415 million pounds in 1977.

In the United States about 200,000 people keep almost 5 million colonies and produce 200 million to 250 million pounds of honey annually. Beekeepers can be classified as full-time, sideline, or hobbyist, with the number of colonies operated by individual owners varying from one to 30,000. Beekeepers derive income from the sale of honey, renting of colonies for crop pollination, production and sale of queen bees and packaged bees, and to a minor extent, from the sale of beeswax, pollen, bee venom, propolis, and royal jelly.

Problems and dangers confront the long-time survival of beekeeping as a profitable agricultural enterprise, and changing agricultural and land-use practices threaten the survival of adequate numbers of bees required to pollinate some 90 crops or more. As human population increases, houses, factories, and highways replace open fields of honey and pollen plants.

Clean cultivation of farmland and large-scale monoculture reduce the sequence of wild plants needed to provide bee food throughout the season.

Pesticides not only kill many bees, but bees also cannot be kept in areas where pesticides are used on a regular basis—such as near fruit orchards and many cottonfields. The presence of nectar and pollen plants in adequate numbers throughout the season is essential to prosperous beekeeping. In the national interest, beekeeping must survive. If it is to do so, it will need greater consideration than it now receives in land-use planning, in the revegetating of disturbed land, in large-scale weed and pest control programs, and in providing beekeeping sanctuaries on State and Federal lands.

Crop pollination is more essential to agricultural production than is generally realized. To maintain an adequate pollinating force of bees in all parts of the country, beekeeping must remain a viable, prosperous industry. Beekeeping will survive in strength adequate to our needs only if we can reverse the trend of recent years toward a deteriorating environment for bees.

The purpose of this handbook is to provide readers with a better understanding of beekeeping in the United States. It is not a beginner's book in the how-to-do-it sense, but it does provide the beginner as well as the experienced beekeeper with a good insight into the status of this small but essential industry.

Dr. E. F. Phillips, one of the early leaders of research in the U.S. Department of Agriculture, said that more had been written about bees than any living thing other than human beings. Books, bulletins, and bee journals still provide evidence that our fascination with bees and beekeeping continues unabated. Beekeepers of this generation must try to make sure that we bequeath an environment in which bees may profitably be kept by future generations.

[1] Retired, formerly staff scientist, National Program Staff, Science and Education Administration.

HISTORY OF BEEKEEPING IN THE UNITED STATES

By EVERETT OERTEL [1]

The honey bee (*Apis mellifera* L.) is not native to the Western Hemisphere. Stingless bees (Meliponids and Trigonids) are native to the West Indies, as well as Central and South America. Wax and small amounts of honey were obtained from stingless bee nests by the early Indians of these areas.

Information available indicates that colonies of honey bees were shipped from England and landed in the Colony of Virginia early in 1622. One or more shipments were made to Massachusetts between 1630 and 1633, others probably between 1633 and 1638. The author was not able to find any records of importing honey bees into other Colonies, but it is reasonable to assume that they were brought by the colonists to New York, Pennsylvania, Carolina, and Georgia.

Records indicate that honey bees were present in the following places on the dates shown: Connecticut, 1644; New York (Long Island), 1670; Pennsylvania, 1698; North Carolina, 1730; Georgia, 1743; Alabama (Mobile), 1773; Mississippi (Natchez), 1770; Kentucky, 1780; Ohio, 1788; and Illinois, 1820 (Oertel *1976*). By 1800, honey bees were widely distributed from the Atlantic Ocean to the Mississippi River.

Honey bees may have been taken to Alaska in 1809 and to California in 1830 by the Russians, according to Pellett (*1938*), but no records are available as to whether they survived. In the 1850's, bees were shipped from the Eastern States to California. A few hives were taken over land, but most of the hives were sent by ship to Panama, by land across the Isthmus, and then by ship to California. Probably, the bees reached Oregon and Washington from California in natural swarms or in hives taken there by settlers. There are no dependable records that describe how bees spread westward from the Mississippi River into the

Mountain States. It seems likely, however, that bees moved into these areas the same way they did into Oregon and Washington; that is, in natural swarms or in hives carried by the early settlers.

Development of Modern Equipment

For thousands of years, colonies of honey bees were kept in wooden boxes, straw skeps, pottery vessels, and other containers. Honeycomb built in such hives could not be removed and manipulated like the movable combs of today. No doubt the first hives used in the American Colonies were straw skeps (fig. 1). Later the abundance of cheap lumber and lack of trained people to make straw hives caused a fairly rapid shift to box hives made of wood. Log gums, that is, sections of bee trees containing colonies of bees, occasionally were

PN-6737

FIGURE 1.—The straw skep was used widely in Europe, but very little in North America.

[1] Retired, formerly apiculturist, U.S. Department of Agriculture.

2

sawed out and used as hives. A few gums may be in use even now, particularly in wooded, isolated areas (figs. 2 and 3). Some ingenious farmers built wood hives with easily removable tops (caps) so that chunks of honey could be removed without killing the colonies. Affleck (*1841*) showed caps (now called supers) in his illustrations, but he did not give any details such as when they were first used.

In 1852, L. L. Langstroth, a Congregational minister from Pennsylvania, patented a hive with movable frames that is still used today. The principle upon which Langstroth based his hive is the space kept open in the hive to allow bees passage between and around combs. This space is about three-eighths of an inch wide; space that is less than this is sealed with propolis and wax, while space wider is filled with comb. Before this time hives were either Greek bar hives or leaf hives that

PN–6739

FIGURE 3.—Bee gum with glass jar on top for honey storage.

allowed the beekeeper to inspect the comb (fig. 4). Langstroth is called "the father of modern beekeeping."

In the period between the importation of honey bees by the early colonist and invention of the movable frame hive by Langstroth, beekeepers had little capability for managing their colonies. They increased their number of colonies each spring by capturing swarms and killed them in the fall by burning sulfur at the entrance of the hive so that the honey and beeswax could be removed. The comb, then, was crushed to squeeze out the honey.

Honey generally was obtained (1) by cutting bee trees and taking what honey was available, (2) by killing colonies and taking the honey within the hive, or (3) by taking whatever honey was stored in a crude super or cap that was placed on the hive during the summer.

Modern methods of beekeeping came very rapidly following Langstroth's patent. Other inventions soon followed that made large-scale, commercial beekeeping possible. Wax-comb foundation,

PN–6738

FIGURE 2.—An unusually tall bee gum.

FIGURE 4.—Book hive with hinged frames used by François Hüber in Switzerland, who published his observations in 1792.

PN-6740

invented in 1857, made possible the consistent production of straight, high-quality combs of predominantly worker cells. Pellett (*1938*) gives a detailed account of the development of wax-comb foundation. The invention of the centrifugal honey extractor in 1865, and its subsequent improvements, made possible large-scale production of extracted honey. The bee smoker, as now used by beekeepers, evolved from a pan used to contain some burning, freely smoking material, the smoke of which could be blown across the open hive to control the bees. The all-important bee veil gradually evolved from pieces of coarse cloth that were wrapped about the head of the beekeeper.

Introduction of Italian Stock

No one knows how many colonies or hives of honey bees were brought to the American Colonies by the first settlers. Nor do we know from what countries they came: England, Holland, France, Spain, or perhaps somewhere else? It is likely that after the early imports all increase was by natural swarming. Since we do not know how many colonies were brought to the east coast, we cannot determine the degree of inbreeding.

In the 1850's, the superior merits of the Italian race of honey bees became known to a few leaders of American beekeeping and they attempted to import queen bees from Italy. Accounts of these first efforts are confusing, but according to Pellett (*1938*), the first known successful importation of Italian queen bees was made in 1860.

During the last part of the 19th century, some queen bees of other races were brought into this country. They were imported from Egypt, Cyprus, the Holy Land, Syria, Hungary, and Tunisia, according to Pellett (*1938*). None of those races, or selections, was of lasting use in the United States, however. Carniolan and Caucasian queen bees also were imported and still are used to a limited extent. The bee journals and the trade catalogs from about 1870 until after World War I carried advertisements for imported queen bees or their progeny, largely Italian stock. Today, the American version of the Italian race is widely used throughout this country.

Imported Italian queen bees were advertised for sale by L. L. Langstroth and Sons, Oxford Ohio, in 1866, but no prices were given. Those interested were advised to write for a price list. In 1867, Adam Grimm, Jefferson, Wis., advertised imported Italian queen bees for sale at $20 each. He promised to sell medium-sized colonies of bees, with imported queens, in movable comb hives for $30 each in 1868. Others who advertised Italian queen bees for sale in 1867 were C. B. Bigelow, Vermont; A. Gray, Ohio; Ellen S. Tupper, Iowa; William W. Cary, Massachusetts; and K. P. Kidder, Vermont. This last group did not quote prices. Egyptian queen bees were offered for sale by Langstroth and Sons and A. Gray, but no prices were quoted. Charles Dadant, Illinois, offered imported Italian queen bees for sale at $12 each.

The originally introduced dark bees of northern Europe predominated throughout much of the United States and Canada during the 1800's and into the 1900's. Strains present toward the end of that era tended to be irritable and nervous, running readily over the combs and hive. These strains were also subject to European foulbrood disease. Queen bees were shipped from Europe in large numbers from the 1880's to 1922, when a law was passed prohibiting further imports. The purpose of this law was to prevent introduction of the acarine mite, which was causing serious problems in Europe, into the United States.

As queen rearing developed into a large-scale commercial enterprise in the Southern States and Italian queens from Europe were used extensively in the breeding program, a strong, Italian-type bee predominated. Before the end of the 1920's, however, after years of persistent requeening with southern queens, northern beekeepers largely replaced the black bees with a less nervous, Italian-type bee that resisted European foulbrood.

Queen Bee Rearing

As the number of colonies owned and operated by individual beekeepers increased, a market developed for young queen bees. In 1861 Henry Alley, William Carey, and E. L. Pratt, all of Massachusetts, began producing queens for sale. These early producers used narrow strips of comb containing eggs and larvae which they fastened to the top bars or partial combs. When these materials were added to swarm boxes that were queenless, queen cells formed. The queen cells were distributed individually to queenless colonies for mating.

G. M. Doolittle, Onondaga, N.Y., in 1889 developed a comprehensive system for rearing queen bees that is the basis of bee production today. His system, essentially, was making wax cups and placing worker bee larvae into them from which the queen-rearing bees formed the queen cells. This same system, or some modification of it, is used today by all commercial queen rearers.

Since 1886 queen bees have been sent in the mail, which has benefited both buyers and sellers (Pellet *1938*). Losses in transit have been reported from time to time, but on the whole, shipment by mail has been satisfactory. Post offices will accept either single queen cages or several cages stapled together. About a million queen bees are sent in the mail annually. Most of these bees are mailed to places in the United States and Canada, but some are sent to other countries.

Recent developments include the crossing of selected inbred lines to produce hybrid bees, and as of 1977, the direct sale of artificially inseminated queens. This step marks the beginning of a new era in bee breeding, in that male and female lines can now be controlled in a commercial breeding program.

Commercial Beekeeping

From the beginning of beekeeping in the 1600's until the early 1800's, we assume that honey was largely an article of local trade. Many farmers and villagers kept a few colonies of bees in box hives to supply their own needs and those of some friends, relatives, and neighbors (fig. 5). According to Pellett (*1938*), Moses Quinby of New York State was the first commercial beekeeper in the United States as his sole means of livelihood was producing and selling honey. Quinby (*1864*) described the

PN-6741

FIGURE 5.—Box hive used widely in the United States before movable frame hives became available.

box hives that he built so that combs of honey could be removed without first killing the colonies. Quinby writes of his financial returns as: "In particularly favorable seasons, hives will yield a profit of one or two hundred percent—in others, they hardly make a return for trouble." Quinby, after experimenting with a few movable comb hives, gradually replaced his box hives with the movable comb-type and advised others to do likewise.

Other beekeepers in Quinby's neighborhood used his methods and began to produce honey on a commercial scale. As the use of movable comb hives, comb foundation, and improved honey extractors became more widespread, commercial beekeeping spread into other States. Poor roads and the use of horse-drawn vehicles restricted the size of the area in which a beekeeper could operate and the number of colonies that could be managed profitably. After World War I, however, with better highways and increased use of motor vehicles and more efficient methods of colony management and honey handling, commercial beekeepers throughout the United States were able to expand the size of their businesses. By 1957 Anderson (*1969*) estimated that 1,200 professional beekeepers operated 1,440,000 colonies in the United States. By that time, hobbyists had a few colonies, the part-time beekeepers kept from 25 to 300 colonies, and the commercial beekeeper had up to several thousand colonies. Some U.S. beekeepers have owned as many as 30,000 colonies.

Comb or Section

The term "section" used here describes the honey produced in small wooden frames or sections. The production of section honey is, to coin a phrase, "the fanciest product of the beekeeper's art." Probably, section honey was first produced in the 1820's. Moses Quinby produced section honey in the 1830's and 1840's and did not claim that the method originated with him. Honey was produced by cutting large holes in the top of a box hive, setting a shallow cap on the hive, and filling the cap with wooden sections that might have small comb starters fastened to them. A cover was placed over the hive. The sections, which were of various sizes, might contain up to 4 pounds of honey when filled. Some beekeepers inverted glass containers over the holes in the box hive, and if they were lucky had honey stored in them.

The crude method of section honey production was gradually abandoned as more and more beekeepers began to use movable comb hives. The large homemade section boxes were replaced with smaller, factory-made ones. Supers especially fitted to hold the sections were developed. Manufacturers sold 45 million to 55 million sections annually in the years just before World War I. Between about 1875 and 1915, approximately one-third of the honey produced in New England, New York, Pennsylvania, the Midwest, and a few Western States was in the form of section honey. Generally, the nectar flow in the Southern States was not suitable for section honey production.

Increase in Production of Extracted Honey

The amount of section honey produced declined rapidly after World War I. The product was fragile and difficult to ship; shelf life was short and combs were likely to leak or granulate. Production of section honey required a heavy nectar flow of several weeks' duration, and a great deal of hand labor for cleaning, weighing, and grading. In addition, beekeepers were unable to provide the intensive colony management needed in outyards miles from their homes. The Pure Food Law of 1906 gave buyers more confidence in the purity of extracted honey, thereby increasing demand for it. During the sugar-short period of World War I, the demand for honey increased and, as the price was high, production of extracted honey increased rapidly.

Large amounts of liquid honey were shipped in wooden barrels in the last part of the 19th century. Then 60-pound metal cans came into general use. Today, most bulk honey is sold in steel drums.

Development of Honey-Packing Plants

As commercial honey producers increased the size of their operations, they found it difficult to pack and sell the crop on the retail market and specialized honey-packing plants developed in the 1920's. Packing plants now are very sophisticated in packing liquid or smoothly crystallized honey.

Beeswax

Beeswax was an article of commerce soon after it became available in the Colonies. It was widely used in candles at home and abroad. The wax was melted, poured into molds, and then transported to market. North Carolina in 1740 and Tennessee in 1785 permitted taxes to be paid in beeswax because of the shortage of money (Oertel *1976*). Information is not available about how much beeswax was produced or used in the Colonies in the 1600's and the first part of the 1700's. Beeswax was an article of export in the 18th century, particularly from the ports of Philadelphia, Charleston, Pensacola, and Mobile. In 1767, a total of 35 barrels of beeswax were exported from Philadelphia and 14,500 pounds from Charleston in 1790. Beeswax was listed in articles exported from the British Continental Colonies in 1770: [2] Value 6,426 pounds sterling; 128,500 pounds weight; 62,800 pounds to Great Britain; 50,500 pounds to Southern Europe; 10,000 pounds to Ireland; and the rest to the West Indies and Africa. Honey was not mentioned.

Bee Supply Manufacturers

No doubt, before the invention of the movable comb hive, beekeepers made their own box hives. Movable comb hives and frames must be cut to exact measurements, so machine methods grad-

---

[2] Taken from *Historical Statistics of the United States*, 2 parts, 1975, Bureau of the Census, U.S. Department of Commerce.

ually took over from manufacture by hand. As metal honey extractors came into general use, companies began to offer them for sale. C. P. Dadant began to sell bee hives and frames to his neighbors in 1863 and comb foundation in 1878.[3] By 1884, Dadant and Sons had sold 60,000 pounds of comb foundation throughout the United States.[3] In 1867, C. B. Bigelow of Vermont advertised that he sold the Langstroth bee hive (fig. 6). In 1868, J. Tomlinson, Wisconsin, had honey boxes and frames for sale. In the same year, the National Bee-Hive Company, Illinois, sold bee hives, frames, honey boxes, and honey extractors.

A. I. Root and Moses Quinby started to sell bee supplies in 1869. In 1870, Henry Alley, Massachusetts, sold the Langstroth hive, and A. V. Conklin, Ohio, sold the Diamond bee hive. Later on in the 1870's, Alley offered the Bay State hive for sale, claiming that this was the "best hive in use." Edward Kretchmer, Iowa, began to manufacture and sell supplies in 1874. The W. T.

---

[3] Personal communication from Dadant & Sons, Inc., Hamilton, Ill.

Falconer Co., New York State, started its bee supply business in 1880. At about this same time, P. L. Viallon, Louisiana, began to manufacture and sell bee hives.

Today's beekeeper, who is used to large colonies of bees, would be amused or puzzled if he could see the small hives used in the American Colonies, and even in the States until about 1900 to 1920. The small hives meant small colonies of bees, small crops of surplus honey, and many swarms. Several old books the author consulted stated that a beekeeper should be well pleased if a colony contained 10,000 to 25,000 bees. Even Moses Quinby, a leading beekeeper in the mid-1880's, stated that a 12- by 12- by 14-inch hive (excluding the cap or super) was large enough for use in New York State and an even smaller hive probably would be adequate in warm climates. Quinby thought that 25 pounds of honey was sufficient to last a colony from October 1 to the following April. Charles Dadant, on the other hand, advocated large hives and strong colonies of bees. Over the years, other beekeepers became convinced that a colony must have a large population at the beginning of the nectar flow, an accepted practice today.

PN-6607

FIGURE 6.—Model of Langstroth's original movable-frame hive, with the front removed to show the frames.

Twentieth Century

During the 20th century, the dimensions of bee hives and frames became more standardized, thus eliminating the various sizes that were so confusing 100 or more years ago. The 10-frame movable comb hive is now used throughout the world wherever beekeeping is seriously practiced. Most beekeepers use full-depth standard hive bodies for brood chambers; some also use them for honey supers, while others use shallow or half-depth bodies. Development of strong colonies for major nectar flows rests upon such fundamentals as hive room, adequate stores, and high-quality queen bees. Commercial and part-time beekeepers control swarming in their colonies, but beginners still have difficulties. Drugs (antibiotics) are now available for the control of foulbrood and nosema disease. Artificial insemination of queen bees, that is, controlled mating, is being used commercially to a limited extent.

The rental of colonies for the pollination of certain crops has increased markedly in this century, although management of colonies for such purposes needs to be improved.

The wax moth (*Galleria mellonella*) has been a serious pest of stored combs and weak hives. A limited survey by Williams (*1976*) showed that in recent years annual losses caused by the wax worm ranged from $48,000 in Louisiana to $1,016,000 in Florida. Such early writers as Affleck (*1841*), Langstroth (*1862*), and Miner (*1859*), gave much space to the damage caused by this pest and how it might be controlled. A number of patents were issued in the 1840's and 1850's for various devices that were supposed to keep wax moths from entering bee hives. None was effective. Chemicals have been used with some success, and the feasibility of using biological control methods is being studied.

Research Sponsored by U.S. Department of Agriculture

A full description of apicultural research, as conducted by the U.S. Department of Agriculture, needs much more space than can be devoted to it here. Consequently, only a brief outline is given. In 1860 William Bruckisch, a German immigrant, suggested that the U.S. Government should conduct investigations in beekeeping, and money was set aside to start such research in 1885. Those who have had responsibility for guiding this program are listed below:

N. W. McLain—1885–87, discontinued because of lack of funds.

Frank Benton—1891–1907, work suspended in 1896–1897; no funds. Spent much of his time locating and shipping stock from Europe.

E. F. Phillips—1905–06, acting; 1907–24

J. I. Hambleton—1924–58

C. L. Farrar—1958–61

F. E. Todd—1961–65

S. E. McGregor—1965–69

M. D. Levin—1969–75

E. C. Martin—1975–79

The following did some of their research while employed in the USDA's Division of Bee Culture. Their names were well known in the earlier part of this century.

James A. Nelson—author of *The Embryology of the Honey Bee.* 1915.

R. E. Snodgrass—author of *Anatomy and Physiology of the Honeybee.* 1925.

G. F. White—basic bulletins on bee disease, 1906–20.

References

AFFLECK, T. H.
1841. BEE BREEDING IN THE WEST. 70 p. E. Lucas, Cincinnati.

ANDERSON, E. D.
1969. AN APPRAISAL OF THE BEEKEEPING INDUSTRY. 38 p. U.S. Department of Agriculture, Agricultural Research Service ARS 41–150. (Series discontinued; Agricultural Research Service is now Science and Education Administration—Agricultural Research.)

BAILEY, M. E., E. A. FIEGER, AND E. OERTEL.
1954. PAPER CHROMATOGRAPHIC ANALYSES OF SOME SOUTHERN NECTARS. *In* Gleanings in bee culture 82(7 and 8).

DOOLITTLE, G. M.
1889. SCIENTIFIC QUEEN-REARING. 169 p. T. G. Newman & Son, Chicago.

LANGSTROTH, L. L.
1862. THE HIVE AND THE HONEY BEE. 409 p. C. M. Saxton, New York.

MINER, T. B.
1859. THE AMERICAN BEEKEEPER'S MANUAL. Ed. 4, 349 p. A. D. Moore & Co., New York.

OERTEL, E.
1976. EARLY RECORDS OF HONEY BEES IN THE EASTERN UNITED STATES. American Bee Journal 116, 5 parts, Feb.–June.

1971. HONEY PRICES SINCE 1880. American Bee Journal 111(2):50–51.

PELLETT, F. C.
    1938. HISTORY OF AMERICAN BEEKEEPING. 393 p. Collegiate Press, Ames, Iowa.

QUINBY, M.
    1864. MYSTERIES OF BEEKEEPING EXPLAINED. Ed. 8, 393 p. C. M. Saxton, New York.

WILLIAMS, J. L.
    1976. STATUS OF THE GREATER WAX MOTH, "GALLERIA MELLONELLA," IN THE UNITED STATES BEEKEEPING INDUSTRY. American Bee Journal 116(11): 524–526.

BEEKEEPING REGIONS IN THE UNITED STATES

By William P. Nye [1]

Based on flora, beekeeping methods, and land topography, the continental United States can be divided into seven geographical regions (fig. 1). Each region is discussed here from the standpoint of honey production and methods of beekeeping operations.

The flora, climate, and nature of the terrain determine the system of management practiced by the beekeeper. For example, in the Appalachicola swamps of the Southeast, hives are placed on scaffolding to protect them from flood waters. In the Southwest, shade must be provided to protect

_____
[1] Retired, formerly apiculturist, U.S. Department of Agriculture.

the hives from the hot sun. Colonies in the north and mountainous areas must be protected from the cold, in certain forested areas from bears, and on the desert from drifting sand.

Beekeepers must pay for some locations; others are furnished free. Where bees are desired for pollination, the beekeepers usually are paid for their services.

Most beekeepers move colonies at night when the bees are all inside the hive. But when daytime temperatures exceed 43.3°C (110°F) in the Southwest, bees stay inside the hive and are more easily moved at midday than at night when they tend to cluster at the entrance.

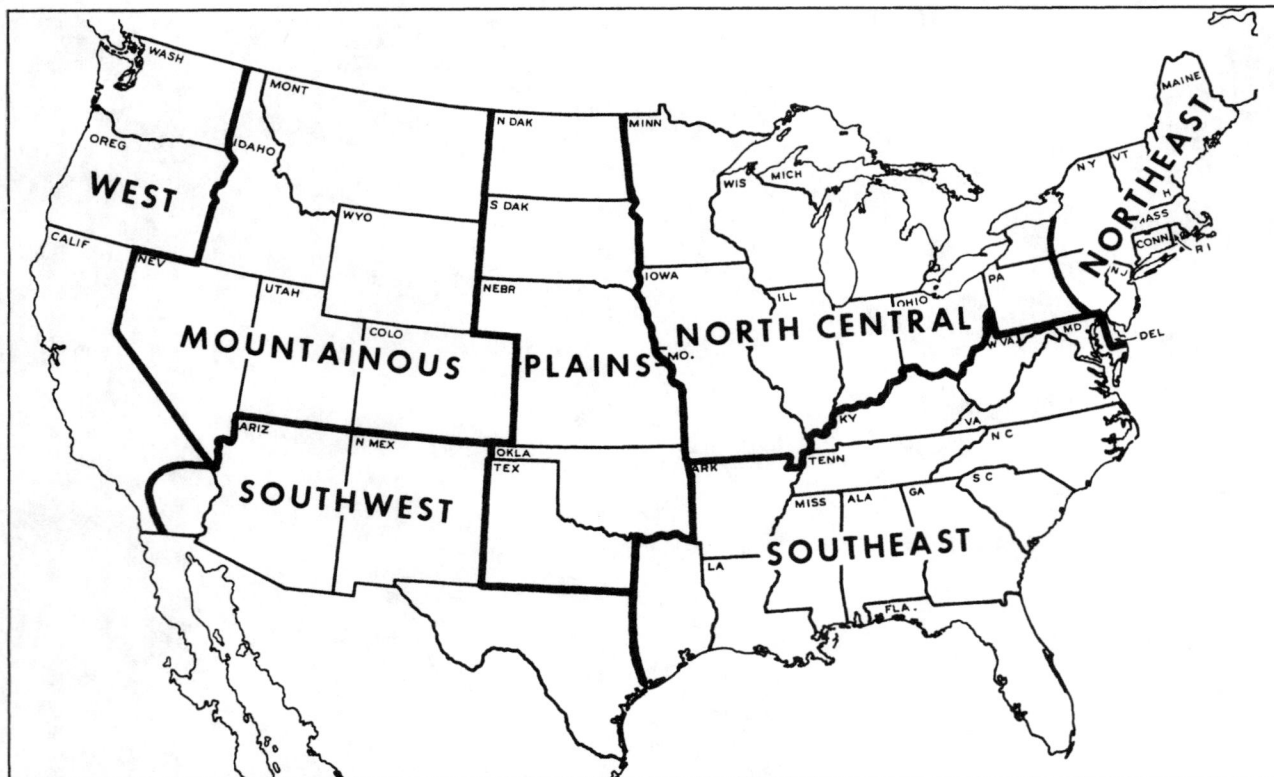

BN-30051

FIGURE 1.—Beekeeping regions of the United States.

Northeast

The severe winters, short summers, and hilly or mountainous nature of the Northeast produce a variety of plants—but none which serves as a major source of nectar. However, alfalfa is becoming an important source of nectar in certain areas as new and better varieties are developed. Nectar from white clover, basswood, black locust, birdsfoot trefoil, various berries, and wild flowers contribute to producing a mixture of honey, much of which is sold locally to residents acquainted with the type produced, and some of the highest prices for honey are obtained here. Few commercial beekeepers operate in the Northeast.

Average honey production per colony is only 13.3 kg (29 lb), but occasionally locations where alfalfa is grown produce much higher averages. An estimated 175,000 colonies are in this region.

The colonies are seldom moved, except the few belonging to commercial or semicommercial beekeepers who may rent their bees for pollination of blueberries, cranberries, other fruits, or cucumbers. Many commercial beekeepers now remove most of the honey, and each hive is reduced to a two-story brood nest that is trucked to the Southeast where it is allowed to build up and be divided to form new colonies. The hives are returned to the Northeast in the spring for fruit pollination before the main honey flow.

Colonies that are not moved South are located where there are good air drainage, protection from cold winds, and exposure to as much winter sun as possible. For additional protection from cold winters, many colonies are "packed," that is, wrapped with insulation and tar paper, leaving only the entrance exposed. Winter loss is usually high and is replaced with package bees and queens purchased from southern beekeepers. Shade in summer is unnecessary.

Most beekeepers overwinter their colonies in two- or three-story, 10-frame standard Langstroth hives. Two basic types of hive covers and bottom boards in use are the telescope cover and reversible bottom board, and the California-style top and bottom. The telescope covers create problems when hives are moved because the hives do not fit closely together on a truck and break open when roped tightly in place. Where migratory beekeeping is practiced, the California-style top and bottom are used as they permit better stacking of hives on a truck. When the honey flow starts, beekeepers add one or two deep supers for surplus honey storage or one or two shallow supers for section or comb honey production.

North-Central Region

The bulk of the honey from the north-central region comes from alfalfa, soybeans, sweetclovers (yellow and white), and the true clovers (alsike, ladino, red, and white), with minor surpluses from basswood, black locust, and raspberry. All of this is high-quality honey. Alfalfa and clover are the predominant American honeys. Less desirable grades come from aster, goldenrod, and smartweed. The variety of other plants, however, ensures something for the bees to work on from spring until frost. The bulk of comb honey produced by bees in 1-pound sections comes from this region.

There are approximately 918,000 colonies, many of which belong to commercial beekeepers. Average production of surplus honey per colony is 24 kg (52 lb).

Some colonies are killed in the fall, and the equipment is stored; then the hives are restocked in the spring with packages of bees and a queen purchased from southern beekeepers. Other colonies are wrapped with insulation and tar paper for winter protection. Some are left with ample stores of honey and pollen in locations protected from wind and exposed to warming sunlight (fig. 2). Still others have most of the honey removed, and the hives are reduced to two-story brood nests that are trucked to the South, where they are allowed to build up and be divided to form new colonies. These are then returned to the North in the spring. Midsummer shade is beneficial. Migratory beekeeping is increasing as beekeepers move their colonies from one location to another to take advantage of the various nectar flows.

Some colonies are rented for pollination of fruits, legumes, and cucumbers.

Southeast

Average production of honey per colony in this region, 14 kg (30 lb), is about the same as in the Northeast but less than elsewhere. An estimated 1,483,000 colonies are located permanently in the Southeast. In addition, many thousands of colonies are trucked in from the northern areas during the winter, then returned to the North in the spring.

Most U.S. queen breeders and package bee shippers are located in the Southeast. An estimated

BN-30066

FIGURE 2.—Apiary sheltered by hardwood forest in north-central region.

300,000 kg (660,000 lb) of live bees and many thousands of queens are shipped from the Southeast annually. Some northern beekeepers pick up their package bees and queens in van-type, air-conditioned trucks for safe transportation to their northern locations.

Except for sizable areas in Florida, little pollination is provided on a cost basis in this region. Bees are rented for occasional pollination of fruit orchards and legume seed and melon production. In Florida, bees are rented for citrus, cucurbits, melons, and other fruits and vegetables.

In the mountainous area, sourwood is the prevailing source of quality honey, along with tulip-poplar and clovers. Sourwood honey is almost water white, does not granulate readily, and is so esteemed that it usually passes directly from producer to consumer at far above the price of other honeys. Various other honeys, from light to dark and from mild to strong, are produced in the Southeast.

In the lower elevations, gallberry becomes the predominant nectar source. In the Appalachicola swamp area, tupelo, famous for its high levulose content and nongranulating characteristics, also is an excellent source of honey. Farther south in Florida, citrus is the major source, with clovers the major source toward the Mississippi Delta, where cotton also becomes important.

Considerable migratory beekeeping occurs, for the long season permits harvest of a crop of honey in one area before another harvest starts elsewhere.

Chunk honey production is common—that is, a chunk of comb honey in a jar of liquid honey. Little section honey is produced.

Preparing bees for winter requires little work. Bees usually are wintered in two- or three-story hives. The problem is to have ample stores of honey and pollen in the colony in the fall. This is necessary for the strong colonies needed in the early spring for package bee production or the early honey flows.

Colonies benefit from shade during the summer in the Southeast, and shade is essential in the southern part for maximum colony production.

Plains Region

The bulk of the honey from the plains region comes from sweetclover and alfalfa; much of it is produced by commercial beekeepers.

In this region, about 476,000 colonies produce 25 kg (55 lb) of honey per colony. Colonies are wintered and operated similarly to colonies in the north-central region. Shade is not generally necessary, although partial or thorough shading during extremely hot midsummer days is beneficial. Some of the highest production per colony is obtained in the plains region. One reason is that the sweetclover and alfalfa fields are relatively large and can support many colonies, and many of the apiaries belong to commercial beekeepers.

Some of the colonies are trucked to southern areas for the winter; some are packed; some are killed and then restocked in the spring; and others receive no special winter treatment.

Colonies are used to a limited extent in the pollination of alfalfa, sweetclover, and cucumbers.

From this region westward to the Pacific, where migratory beekeeping is practiced to a greater extent than elsewhere, the California-style top and bottom rather than the telescoping top and reversible bottom are used, as they permit better stacking of colonies on trucks.

Mountain Region

The major source of honey in the mountainous region is alfalfa (figs. 5 and 6). About 330,000 colonies produce on an average 30 kg (66 lb) of honey per colony. More than two-thirds of the colonies belong to commercial beekeepers; some manage 2,000 colonies or more with only part-time summer help.

Honey production is almost entirely dependent on irrigation, although alfalfa now is grown on dry land. Some of the highest production per colony is obtained in the mountainous region. One reason for this is that the alfalfa fields are relatively large and can support many colonies (fig. 3). Weed spraying has reduced the sweetclover acreage, but sweetclover is now on the increase in some areas.

In migratory beekeeping from this area west and south, the colonies usually are moved at night. The hive entrances are not closed, but the truckload usually is covered with a plastic screen for long trips. Some colonies are packed during the winter, which is extremely cold and dry. Colonies not packed are located where they have good wind protection, exposure to the sun, and good air drainage. Spring buildup is slow and fall nectar flows are rare. Shading is unnecessary.

Migratory beekeeping is extensive. For example, probably no other region in the country can com-

BN-30056

FIGURE 3.—Unsheltered colonies located for alfalfa honey production and pollination in Utah.

pare with the Delta area of central Utah with so many colonies (20,000 to 40,000) moved in from such long distances in so short a period. The region produces a major portion of the alfalfa seed in Utah (fig. 4). Many colonies are moved south or west for the badly needed spring buildup, then returned for the summer flow. Some colonies are killed in the fall and restocked in the spring.

Southwest

In this hot, semiarid region, there are 155,000 colonies that produce 21.4 kg (47 lb) of honey per colony. The major sources of nectar are alfalfa, cotton, and mesquite. Other sources include citrus, catclaw, tamarix, safflower, wild buckwheat, and other desert shrubs.

Summer shade is highly important (fig. 5). Artificial shade is often provided. Winter protection is unnecessary. Some colonies are wintered in a single brood nest with one or two shallow supers, but most are in two or three standard hive bodies. Nearby water is essential, and if it disappears even for only a day, the colonies may perish. Migration from one honey flow to another is common.

PN-6742

FIGURE 4.—Colonies in groups of 8 to 12 are placed one tenth mile apart in large alfalfa seed fields for pollination

BN-30071

FIGURE 5.—Typical apiary under a ramada that partially shades colonies in hot Southwest.

Colonies are used extensively in pollination of alfalfa and melons and to a lesser degree for citrus, onions, and cotton. A few package bees and queens are produced, but for the most part bees are kept for production of honey by commercial operators. Apiaries of 100 colonies or more are not unusual.

West

About 668,000 colonies in this region produce 12 kg (27 lb) of honey per colony. This production is rather meaningless because of the differences due to extreme variations in temperature, rainfall, elevation, and flora. The main source of nectar is alfalfa, which produces a light-colored honey of excellent flavor. Cultivated field crops such as clover, citrus, cotton, lima beans, deciduous fruit trees, and cucurbits are important sources of pollen and nectar during their blooming periods.

Other plants such as wild buckwheat, star thistle, sage, and fireweed in restricted localities may yield commercial quantities of honey in favorable years and may rank high in the estimation of the beekeeper because of their value to the bees as a source of food for building up the colony early in the spring or to carry it over the winter period.

The region varies in rainfall from 1 to 2 inches in the desert areas to more than 60 inches in the rain-forest area, in elevation from below sea level to snow-capped mountains, and in temperatures from dry and hot to humid and extremely cold.

Migratory beekeeping is practiced by most of the commercial beekeepers, and four or more moves per year are not uncommon. In California, most of the bees are held in almond areas during the winter. The almond orchards are distributed from Chico in Butt County in the north to Kern County in the south. The pollination season begins with almond blooms in early February. As the almonds finish blooming, the plums and prunes begin to bloom and the beekeeper may move to these. Cherries bloom near mid-March through mid-April. After this period of fruit bloom, there is a dearth of pollen and nectar in cultivated areas.

To maintain and build up colonies for summer pollination service, the beekeeper moves his bees to the mountains where bees are held in manzanita and sage at elevations around 2,000 to 6,000 feet.

Native plants supply pollen and nectar in the Sierra Nevada range, the coast ranges, and coastal areas between the Pacific Ocean and the coast ranges.

Commercial pollination service begins again in June–July with melon pollination, ladino clover, and alfalfa seed production. In the fall, after these sources have been harvested, the beekeeper moves his bees into native flora along the east side of the coast range. This is a major source of nectar and pollen for winter stores from August to frost.

California beekeepers south of the Tehachapi Mountains begin to build up their colonies on native plants in January. Until citrus bloom in April, this is the main bee pasture. Some southern California beekeepers move into the southern and central almond areas in February and March and into alfalfa and cotton during the summer and early fall.

It is evident, therefore, that the beekeeper must move his bees to take advantage of pastures offered by native and cultivated plants during the period. The placement of 2,000 colonies from several beekeepers in a solid square mile of alfalfa grown for seed is not unusual. The use of bees for pollination is extensive. An estimated one-half or more of all colonies are used sometime during the year for pollination hire.

In the last few years, many beekeepers have had to replace almost 100 percent of their colonies due to pesticide losses. These losses are increasing each year. The major dollar loss to beekeeping in California is caused by (1) pesticides, (2) wax moth, and (3) foulbrood diseases.

Beekeepers operate an average of 2,000 colonies. In such operations, the apiary rather than the colony is considered a unit. Such manipulations as requeening, supering, and removing honey are performed on all colonies regardless of their relative condition. Each year, more than 272,727 kg (600,000 lb) of bees and approximately 400,000 queens are shipped from the West.

NECTAR AND POLLEN PLANTS

By Everett Oertel [1]

A beekeeper must have available data on the nectar and pollen plants in the vicinity of his apiary for successful honey production. Such information enables him to determine when to install package bees, divide colonies, put on supers, use swarm-control measures, remove honey, requeen, prepare colonies for winter, and locate profitable apiary sites.

Vansell (*1931*) listed 150 species of nectar and pollen plants in California, but only six are principal sources for commercial honey production. He listed about 90 species of nectar and pollen plants in Utah but noted that the main sources of commercial honey are alfalfa and sweetclover (*1949*). Wilson et al. (*1958*) observed honey bees visiting the blossoms or extrafloral nectaries of 110 species of plants in Colorado, of which the most important honey sources were alfalfa, yellow sweetclover, and dandelion (fig. 1).

PN-6743

FIGURE 1.—Honey bee on a white clover blossom.

---

[1] Retired, formerly apiculturist, U.S. Department of Agriculture.

Beekeepers are advised to record the blossoming period for the nectar and pollen plants in their vicinity (fig. 1). Most State agricultural extension services have publications available on beekeeping. These publications usually contain a list of the important nectar and pollen plants. Unknown plants can be sent to the botany department of the State university for identification

Beekeeping Locations

Beekeepers, especially commercial operators, have learned that the nectar- and pollen-producing plants may change considerably over the years. Variations may be caused by droughts, changes in agricultural crops and practices, irrigation projects, and subdivision development. Changes have been particularly rapid since World War II and are likely to continue.

Acreages planted to buckwheat, alsike clover, and cotton have decreased, whereas those with alfalfa hay, mustard, safflower, and soybeans have increased. In some States, certain soybean varieties are valuable sources of nectar.

Other changes in agricultural practices include the use of herbicides and power mowing machines. They reduce or eliminate plants that are important sources of nectar or pollen. Farmers are depending less on legumes to add nitrogen to the soil and are using more fertilizer.

Pests, such as insects or nematodes, may cause so much damage to some plant species that farmers change to other crops. In Ohio in 1966, the alfalfa weevil (*Hypera postica* (Gyllenhal)) became so destructive that there was serious concern farmers would stop growing alfalfa, an important nectar source. The sweetclover weevil (*Sitona cylindricollis* Fahraeus) has destroyed much of the sweetclover that formerly was grown in the Midwest. The acreage planted for seed has decreased over 50 percent since 1950. Plant breeders have introduced a nectarless cotton so that destructive insects will not be attracted to the plant.

16

No information is available on the effects of air pollution caused by factories, motor vehicles, radioactivity, and major metropolitan areas on nectar and pollen plants except in limited areas. Productive locations for the commercial beekeeper will become more difficult to find.

Poisonous Honey Plants

Fortunately, the American beekeeper is seldom concerned about plants that are poisonous to honey bees. Locations with abundant growth of California buckeye (*Aesculus* spp.), deathcamas (*Zigadenus venenosus*), locoweed (*Astragalus* or *Oxytropis* spp.), laurel (*Kalmia* sp.), or rhododendron (*Rhododendron* spp.) should be avoided, if possible, while these plants are in bloom. Damage to colonies from poisonous nectar or pollen may be severe in some years, but of small consequence in others.

Nectar Secretion

Beginners in beekeeping frequently ask: "Are there any plants that I can grow that will increase my yield of honey?" In general, growing a crop for the bees alone is economically impractical. Beekeepers are dependent on cultivated crops grown for other purposes or on plants growing wild. Certain nectar and pollen plants, such as alfalfa, the clovers, and sweetclover, are grown widely for agricultural purposes and they are wild to some extent. These plants, together with others, such as citrus (orange, grapefruit, lemon, limes, tangelos), cotton, sage and tupelo, furnish the greater part of the Nation's commercial honey.[2]

Sometimes, friendly farmers will seed small areas near an apiary with nectar–producing species, if the beekeeper provides the seed, and thus honey production increases. A few ornamental flowers or trees on a city lot are of small value to an apiary or a colony of bees. Up to several acres of abundant flowers are usually necessary to provide sufficient nectar for one colony (Oertel *1958*).

Nectar secretion or production is affected by such environmental factors as soil type, soil condition, altitude, latitude, length of day, light

[2] The reader who wishes to read a detailed account of the production of nectar is referred to the chapter by R. W. Shue in *The Hive and the Honey Bee*, 1975, Dadant & Sons, Hamilton, Ill. 740 p.

conditions, and weather. Such soil conditions as fertility, moisture, and acidity may affect not only the growth of the plant but also the secretion of nectar. Luxuriant plant growth does not necessarily imply that maximum nectar secretion will take place. At times, limited growth results in increased nectar production. Clear, warm, windless days are likely to favor nectar secretion. Most of our information on nectar production is based only on casual observation.

Nectar is secreted by an area of special cells in the flowers called a nectary. Certain species, such as vetch, cotton, partridgepea, and cowpeas, produce nectar from tiny specialized areas in the leaves or stems called extrafloral nectaries.

Honeydew

Honeydew is the sweet liquid secreted by certain insects, such as aphids or plant lice, scale insects, gall insects, and leafhoppers, and also by the leaves of certain plants. Honeydew honey differs chiefly from floral honey in its higher dextrin and mineral content. The quality of honeydew honey varies greatly. Some types are fairly palatable, whereas others are undesirable for human food or for wintering bees in northern areas.

Pollen Plants

Pollen is an essential food used in the rearing of honey bee larvae and maturing of young worker bees. A good, strong colony of bees may collect and use 50 to 100 pounds of pollen during the season. Lack of pollen slows colony development in many localities in the spring and in some locations in the summer and fall. Pollen may be available in the field, but cold or rainy weather may prevent the bees from gathering it. Some beekeepers feed pollen supplements, alone or mixed with bee-gathered pollen, to their colonies. Pollen supplements are sold by bee-supply dealers.

Nectar and Pollen Plant Regions

In table 1 the nectar and pollen plants are listed by region (fig. 1, p. 16). Some species are limited to a small area within a region; for example, thyme in New York, fireweed in the North and West, gallberry in the Southeast, and citrus in the Southeast, Southwest, and West.

Table 1.—*Nectar and pollen plants by regions*

Plant	Northeast	Northcentral region	Southeast [1]	Plains region	Mountainous region [2]	Southwest	West [3]	Alaska [4]	Hawaii [5]
Alder (*Alnus* spp.)						X	X	X	
Alfalfa (*Medicago sativa* L.)	X	X		X	X	X	X	X	
Algaroba (*Prosopis chilensis* (Mol.) Stuntz)									X
Alkaliweed (*Hemizonia* spp.)						X			
Almond (*Prunus amygdalus* Batsch.)						X			
Amsinckia (*Amsinckia* spp.)						X			
Ash (*Fraxinus* spp.)					X	X			
Aster (*Aster* spp.)	X	X	X	X		X	X		
Baccharis (*Baccharis* spp.)						X			
Balsamroot (*Balsamorrhiza* spp.)					X				
Basswood (*Tilia americana* L.)	X	X	X	X	X				
Bermudagrass (*Cynodon dactylon* (L.) Pers.)						X	X		
Bindweed (*Convolvulus* spp.)					X	X			
Birdsfoot trefoil (*Lotus conriculatus* L.)	X						X		
Bitterweed (*Helenium amarum* (Raf.) Rock)			X						
Blackberry (*Rubus* spp.)	X		X				X		X
Black wattle (*Acacia* spp.)									X
Bladderpod (*Lesquerella gordonii* (Gray) Wats.)						X			
Blueberry (*Vaccinium* spp.)	X	X						X	
Bluecurls (*Truchistema* spp.)						X			
Blue thistle (*Echium vulgare* L.)	X		X						
Bluevine (*Gonolobus laevis* Michx.)		X							
Blueweed (*Cichorium intybus* L.)							X		
Boneset (*Eupatorium* spp.)			X	X				X	X
Boxelder (*Acer* spp.)					X				
Broomweed (*Gutierrezia texana* (DC.) T. & G.)				X	X	X			
Buckbrush (*Symphoricarpos* spp.)		X		X		X			
Buckeye (*Aesculus californica* (Spach) Nutt)							X		
Buckthorn (*Rhamnus* spp.)	X		X	X		X	X		
Buckwheat (*Fagopyrum esculentum* Moench)		X	X						
Burroweed (*Haplopappus tenuisectus* (Greene) Blake ex Benson)					X	X			
Button bush (*Cephalanthus occidentalis* L.)	X	X	X	X			X		
Cacti (*Cactaceae* family)						X			X
Camphorweed (*Heterotheca subaxillaris* (Lam.) Britt. & Lusby)					X	X			
Cascara (*Rhamnus purshiana* DC.)							X		
Catclaw (*Acacia greggii* Gray)						X			
Catnip (*Nepta cataria* L.)	X	X							
Ceanothus (*Ceanothus* spp.)					X	X			
Cedar elm (September elm) (*Ulmus serotina* Sarg.)				X					
Chicory (*Chichorium intybus* L.)	X	X	X	X	X	X	X		
Chinese tallow tree (*Sapium sebiferum* L.)			X			X			

See footnotes at end of table.

TABLE 1.—*Nectar and pollen plants by regions*—Continued

Plant	North-east	North-central region	South-east [1]	Plains region	Mountainous region [2]	South-west	West [3]	Alaska [4]	Hawaii [5]
Citrus (*Citrus* spp.)			X			X	X		X
Cleome (*Cleome serrulata* Pursh)					X	X	X		
Clethra (*Clethra alnifolia* L.)	X								
Clover:									
Alsike (*Trifolium hybridum* L.)	X	X			X		X	X	
Crimson (*Trifolium incarnatum* L.)			X						
Persian (*Trifolium resupinatum* L.)	X	X	X	X					
Red (*Trifolium pratense* L.)	X	X		X	X		X	X	X
Sweetclover (*Melilotus* spp.)	X	X	X	X	X	X	X		X
White (*Trifolium repens* L.)	X	X	X	X	X		X	X	X
Coffee (*Coffea arabica* L.)									X
Cone flower (*Rudbeckia* spp.)	X	X	X						
Corn (*Zea mays* L.)		X	X	X	X	X	X		
Cotton (*Gossypium* spp.)			X	X		X	X		
Cottonwood (*Populus* spp.)			X		X	X	X		
Cowpea (*Vigna sinensis* (Torner) Savi)			X						
Cranberry (*Vaccinium macrocarpon* Ait.)	X	X					X		
Creosote bush (*Larrea tridentata* (DC.) Coville)						X	X		
Crownbeard (*Verbesina* spp.)			X	X		X			X
Cucurbits:									
Cantaloup (*Cucumis melo* L.)	X	X	X	X		X	X		
Cucumber (*Cucumis* spp.)	X	X	X	X	X	X	X		
Gourds (*Cucurbita* spp.)	X	X	X						
Melon (*Citrullus* spp.)	X	X	X			X	X		
Pumpkin (*Cucurbita* spp.)	X	X	X		X		X		
Squash (*Cucurbita* spp.)	X	X	X				X		
Dandelion (*Taraxacum* spp.)	X	X			X		X	X	X
Dogbane (*Apocynum androsaemifolium* L.)	X	X	X						
Eardropvine (*Brunnichia cirrhosa* Gaertn.)			X						
Elm (*Ulmus* spp.)	X	X	X	X	X	X			
Eucalyptus (*Eucalyptus* spp.)						X	X		X
Filaree (*Erodium* spp.)						X	X		
Fireweed (*Epilobium angustifolium* L.)		X					X		
Fruit bloom:									
Apple (*Malus* spp.)	X	X	X	X	X		X	X	
Apricot (*Prunus* spp.)					X	X	X		
Cherry (*Prunus* spp.)	X	X			X		X		
Citrus (*Citrus* spp.)			X			X	X		X
Peach (*Prunus* spp.)	X	X	X	X	X	X	X		
Pear (*Pyrus* spp.)	X	X	X	X	X		X		
Plum (*Prunus* spp.)	X	X	X	X	X		X		
Gallberry (*Ilex glabra* (L.) Gray)			X						
Giant hyssop (*Agastache foeniculum* (Pursh) Ktse.)	X	X							
Goldenrod (*Solidago* spp.)	X	X	X	X	X	X			X
Grape (*Vitis* spp.)			X						

See footnotes at end of table.

TABLE 1.—*Nectar and pollen plants by regions*—Continued

Plant	North-east	North-central region	South-east [1]	Plains region	Moun-tainous region [2]	South-west	West [3]	Alaska [4]	Hawaii [5]
Greasewood (*Sarcobatus vermiculatus* (Hook.) Torr.)					X				
Guajillo (*Acacia berlandieri* Benth.)						X			X
Guava (*Psidium guajava* L.)									
Gumweek (*Grindelia* spp.)				X	X	X			
Hemp (*Cannabis sativa* L.)							X		
Holly (*Ilex opaca* Ait.)			X						
Horsemint (*Monarda* spp.)				X		X			
Huckleberry (*Gaylussacia* spp.)			X						
Hue (*Lagenaria siceraria* (Mol.) Standley)									X
Ilima (*Sida* spp.)									X
Jackass clover (*Wislizenia refracta*)						X	X		
Johnsongrass (*Sorghum halepense* (L.) Pers.)						X	X		
Kly (*Acacia* spp.)									X
Knapweed (*Centaurea repens* L.)					X				
Koa haole (*Acacia* spp.)									X
Lantana (*Lantana* spp.)									X
Laurel cherry (*Prunus caroliniana* Mill.)	X	X	X						
Lima beans (*Phaseolus limensis* Macf.)			X				X		
Locoweed (*Oxytropis* or *Astragalus* spp.)					X	X			
Locust:									
Black (*Robinia pseudo-acacia* L.)	X	X	X	X	X	X			
Thorny (*Gleditsia triacanthos* L.)			X						
Water (*Gleditsia aquatica* Marsh)			X						
Looses rife (*Lythrum* spp.)	X	X							
Lupine (*Lupinus* spp.)						X	X		
Macadamia (*Macadamia* spp.)									X
Mamane (*Sophora* spp.)									X
Mangrove:									
Black (*Avicennia nitida* Jacq.)			X						
Red (*Rhizophora mangle* L.)			X						
White (*Laguncularia racemosa* (L.) Gaertn. F.)			X						
Manzanita (*Arctostaphylos* spp.)					X	X	X		
Maple (*Acer* spp.)	X	X	X	X	X		X		
Matchweed (*Gutierrezia sarathrae* (Pursh) Britt. & Rusby)					X	X			
Mesquite (*Prosopis juliflora* (SW.) DC.)						X	X		
Mexican clover (*Richardia scabra* L.)			X						
Milkvetch (*Astragalus* spp.)					X				
Milkweed (*Asclepias* spp.)	X	X	X	X	X				
Mint (*Mentha* spp.)		X	X				X		
Monkeypod (*Samanea* spp.)									X
Mountain apple (*Eugenia malaccensis* L.)									X
Mule ear (*Wyethia* spp.)					X				
Mustard (*Brassica* spp.)		X			X	X	X	X	
Nohu (*Tribulus cistoides* L.)									X
Oak (*Quercus* spp.)			X		X	X	X		

See footnotes at end of table.

TABLE 1.—*Nectar and pollen plants by regions*—Continued

Plant	North-east	North-central region	South-east[1]	Plains region	Moun-tainous region[2]	South-west	West[3]	Alaska[4]	Hawaii[5]
Ohia lenhua (*Metrosideros* spp.)									X
Oi (*Verbena* spp.)									X
Oregon grape (*Berberis nervosa* Pursh)							X	-------	
Oregon maple (*Acer macrophyllum* Pursh)							X	-------	
Paintbrush (*Castilleja* spp.)									X
Palmetto (*Sabel* spp.)			X						
Palmetto, saw (*Serenoa repens* (Bartr.) Small)			X						
Palm trees (*Palmaceae* family)									X
Partridgepea (*Chamaecrista* spp.)			X						
Pepperbush (*Clethra alinfolia* L.)			X						
Peppervine (*Ampelopsis arborea* (L.) Koehne)			X	X					
Persimmon (*Diospyros virginiana* L.)			X	X					
Pili (*Heteropogon contortus* (L.) Beauv. ex Roem. & Schutt.)									X
Pine (*Pinus* spp.)							X	-------	
Pluchea (*Pluchea* spp.)									X
Poison ivy (*Rhus* spp.)	X	X	X					X	
Poison oak (*Rhus* spp.)	X	X	X					X	
Privet (*Ligustrum* spp.)	X	X	X		X		-------		
Poplar (*Populus* spp.)	X	X	X	X	X		-------	X	
Rabbitbrush (*Chrysothamnus* spp.)					X	X	-------		
Ragweed (*Ambrosia* spp.)		X	X	X	X	X	-------		
Rape (*Brassica napus* L.)				X					
Raspberry (*Rubus* spp.)	X	X	X				X	X	
Rattenvine (*Berchemia scandens* (Hill) K. Koch)			X	X					
Redbud (*Cercis canadensis* L.)			X	X					
Resinweed (*Grindelia* spp.)				X	-------	X			
Russian-thistle (*Salsola* spp.)				X	X	X	X	-------	
Safflower (*Carthamus tinctorius* L.)				-------	X	X	X	-------	
Sage (*Salvia* spp.)	X	X	X	X			X	-------	
Saguaro (*Carnegiea gigantea* (Engelm.) Britt. & Rose)						X	-------		
Sainfoin (*Onobrychis* spp.)					X		-------		
Saltcedar (*Tamarix gallica* L.)					X	X	X	-------	
Santa maria (*Parthenium hysterophorus* L.)			X						
Silky oak (see silver oak)									X
Silver oak (*Grevillea robusta* A. Cunn.)									X
Smartweed (*Polygonum* spp.)	X	X	X	X	-------	X	X	-------	
Snakeweed (see matchweed)					X	X		-------	
Snowberry (*Symphoricarpos occidentalis* L.)	X	X	X						
Snowvine (*Mikania scandens* (L.) Willd.)			X						
Sorghum (*Sorghum* spp.)				X	-------	X	X	-------	
Sourwood (*Oxydendrum arboreum* (L.) DC.)			X						
Soybeans (*Glycine max* (L.) Merr.)		X	X						
Spanish-needles (*Bidens bipinnata* L.)		X	X	X					
Star thistle (*Centaurea solstitialis* L.)							X	-------	

See footnotes at end of table.

TABLE 1.—*Nectar and pollen plants by regions*—Continued

Plant	North-east	North-central region	South-east [1]	Plains region	Moun-tainous region [2]	South-west	West [3]	Alaska [4]	Hawaii [5]
Star thistle (*Centaurea maculosa*)		X					X		
Sumac (*Rhus* spp.)	X	X	X	X					
Summer farewell (*Petalostemun* spp.)			X						
Sunflower (*Helianthus* spp.)			X	X		X			
Tamarix (*Tamarix aphylla* (L.) Karst.) (*Tamarix articulata* Vahl)						X	X		
Tarweed (*Hemizonia* spp.)							X		
Thistle (*Sonchus arvensis* L. and *Cirsium* spp.)			X	X					
Canadian (*Cirsium arvense* (L.) Scop.)							X		
Thyme (*Thymu* sp.)	X								
Tievine (*Convolvulus* or *Ipomoea* spp.)			X	X					
Titi:									
Black (*Cliftonia monophylla* (Lam.) Britton ex Sarg.)			X						
Spring (*Cyrilla racemiflora* L.)			X						
Summer (*Cyrilla* spp.)			X						
Toyon (*Photinia arbutifolia* Lindl.)							X		
Tulip poplar (*Liriodendron tulipifera* L.)	X	X	X						
Tupelo (*Nyssa* spp.)			X						
Vervain (*Verbena* spp.)			X						
Vetch (*Vicia* spp.)			X	X	X		X		X
Vine maple (*Acer circinatum* Pursh)							X		
Wild alfalfa (*Lotus* spp.)							X		
Wild buckwheat (*Eriogonum* spp.)					X	X	X		
Wild currants (*Ribes* spp.)			X		X				
Wild dandelion (*Hymenopappus arenosus* Heller)					X				
Wild snowberry (*Symphoricarpos* spp.)					X				
Willow (*Salix* spp.)	X	X	X	X			X	X	
Wingstem (*Actinomeris alternifolia* (L.) DC.)			X						
Yellow ginger (*Hedychium flavescens* Carey)									X
Yellow-rocket (*Barbarea vulgaris* R. Br.)	X	X							

[1] Morton, J. F. Honeybee Plants of South Florida. 1964. 77th Proceedings of the Florida State Horticultural Society, p. 415–36.

[2] Wilson, W. T., J. O. Moffett, and H. D. Harrington. 1958. Nectar and Pollen Plants of Colorado. Colorado Agricultural Experiment · Station Bulletin 503–S, 72 p.; Vansell, G. H. Pollen and Nectar Plants of Utah. 1949. Utah Agricultural Experiment Station Circular 124, 28 p.

[3] Vansell, G. H. Nectar and Pollen Plants of California. 1931. California Agricultural Experiment Station Bulletin 517, 55 p.

[4] Washburn, R. H. Beekeeping in the Land of the Midnight Sun. 1961. Gleanings in Bee Culture 89: 720–723, 756.

[5] Botanical names taken from M. C. Neal, Gardens of Hawaii. 1965. (Honolulu) Bishop Museum Special Publication 50, 924 p.; nectar and pollen names taken from E. J. Dyce, Beekeeping in the 50th State. 1959; Gleanings in Bee Culture 87: 647–651; and J. E. Eckert and H. A. Bess, Fundamentals of Beekeeping in Hawaii. 1952. Hawaii University Bulletin 35, 32 p.

References

ABRAMS, L., and R. S. FERRES.
  1960. ILLUSTRATED FLORA OF THE PACIFIC STATES, WASHINGTON, OREGON, AND CALIFORNIA. 4 vol. Stanford University Press, Stanford, Calif.

BAILEY, L. H.
  1949. MANUAL OF CULTIVATED PLANTS. 1116 p., revised. Macmillan Co., New York.

BLAKE, S. F.
  1954. GUIDE TO POPULAR FLORAS OF THE UNITED STATES AND ALASKA. 56 p. U.S. Department Agriculture Bibliography Bulletin 23.

FERNALD, M. L.
  1950. GRAY'S MANUAL OF BOTANY. Ed. 8, 1632 p. American Book Co., New York.

JEPSON, W. L.
  1957. MANUAL OF THE FLOWERING PLANTS OF CALIFORNIA. 128 p. California University Press, Berkeley.

LOVELL, H. B.
  1956. HONEY PLANTS MANUAL. 64 p. A. I. Root Co., Medina, Ohio.

NEAL, M. C.
  1965. IN GARDENS OF HAWAII. 924 p. (Honolulu) Bishop Museum Special Publication 50.

OERTEL, E.
  1958. NECTAR YIELDS OF VARIOUS PLANT SPECIES AT BATON ROUGE IN 1955. Proceedings of 10th International Congress on Entomology, No. 4, p. 1027–1029.

PELLET, F. C.
  1976. AMERICAN HONEY PLANTS. 467 p. Dadant & Sons. Hamilton, Ill.

SMALL, J. K.
  1933. MANUAL OF THE SOUTHEASTERN FLORA. 1554 p. Science Press Printing Co., Lancaster, Pa.

VANSELL, G. H.
  1931. NECTAR AND POLLEN PLANTS OF CALIFORNIA. California Agricultural Experiment Station Bulletin 517, 55 p.

———
  1949. POLLEN AND NECTAR PLANTS OF UTAH. Utah Agricultural Experiment Station Circular 124, 28 p.

VINES, R. A.
  1960. TREES, SHRUBS AND WOODY VINES OF THE SOUTHWEST. 1104 p. Texas University Press, Austin.

WILSON, W. T., J. O. MOFFETT, and H. D. HARRINGTON.
  1958. NECTAR AND POLLEN PLANTS OF COLORADO. Colorado Agricultural Experiment Station Bulletin 503–S, 72 p.

HONEY BEE LIFE HISTORY

By Gordon D. Waller [1]

A honey bee begins life as an egg. Bee eggs develop in the ovarioles or small tubes that make up the two ovaries of a queen. The egg is nourished and grows as it moves down this tube. When it is fully formed, it reaches the end of the ovariole, then moves through the oviducts into the vagina. The sex of the new bee is normally determined as the egg passes through the vagina. A lifetime supply of sperm (5 million to 6 million) is stored by each queen in the spermatheca, a little globular sac attached to the vagina. The queen controls the release of sperm with the so-called sperm pump. If an egg is fertilized, it will develop into a female bee, but if not fertilized, a male bee will result. The result is that male bees have only one set of chromosomes (haploid) acquired from the queen.

The queen bee attaches each egg to the base of an empty cell in combs that have been cleaned by workers. The honey bee egg is a smooth, white, sausage-shaped object about 1.5 ml in length. During the first day, the egg nucleus divides—if the egg is unfertilized; or if the egg is fertilized, the fusion nucleus or zygote divides. It is not until the third day that the embryo form (with head and body segments) can be seen within the egg. The head is present at the larger unattached end and the back (dorsum) is on the in-curved (concave) side.

The first sign of hatching occurs when an egg is 72 to 84 hours old. Muscular contractions by the embryo cause a gentle, weaving motion that apparently results in a tiny hole being torn in the outer membrane (chorion). Fluid from within the egg soon emerges and covers the external surface. The embryo with its "tail" attached to the base of the cell continues to move about until its head also touches the base and an arch is formed. In this "croquet wicket" stage, the chorion evidently is dissolved. The larva then eases itself over against

[1] Entomologist, Science and Education Administration, Carl Hayden Center for Bee Research, Tucson, Ariz. 85719.

the bottom of the cell into the familiar C-shaped position (fig. 1).

Honey bee larvae are fed a nutritious substance called royal jelly secreted by the brood-food glands (hypopharyngeal glands) of young workers. During the first 24 hours, worker larvae are fed lavish amounts of royal jelly by older nurse bees. During the second 24 hours, they get very little additional food and thereafter are cared for by nurse bees of all ages. Pollen and honey are present in the food of older worker larvae.

Honey bees use two systems of feeding larvae. Young larvae are fed amounts excessive to their needs and older larvae are provided small quantities of food as needed. It has been estimated that 110,000 visits are made to a single bee during its egg and larval stages, 3,500 of these during the last 24 hours.

A female larva fed continuously on lavish amounts of royal jelly and provided a large, peanut-shaped cell will become a queen. Another larva given a mixture of honey and pollen during the latter half of its larval life and kept in a worker cell becomes a worker. The process that produces the complete expression of sexual characteristics in a queen has not been determined; however, it is considered to be caused by differences in both the quality and the quantity of the larval food provided.

Drone larvae grow larger than either workers or queens and, therefore, require more food. Food given to young drone larvae is nearly devoid of pollen and is milky-white, while that given to older drone larvae is a yellow-brown color and contains considerable pollen. The food given older drone larvae also is higher in pollen content than that given older worker larvae. Thus, both qualitative and quantitative differences distinguish the larval food given queen, worker, and drone.

The developing honey bee larva is a helpless creature whose principal function is eating. Both the malpighian tubules (analogous to human kidneys) and midgut are shut off from the intes-

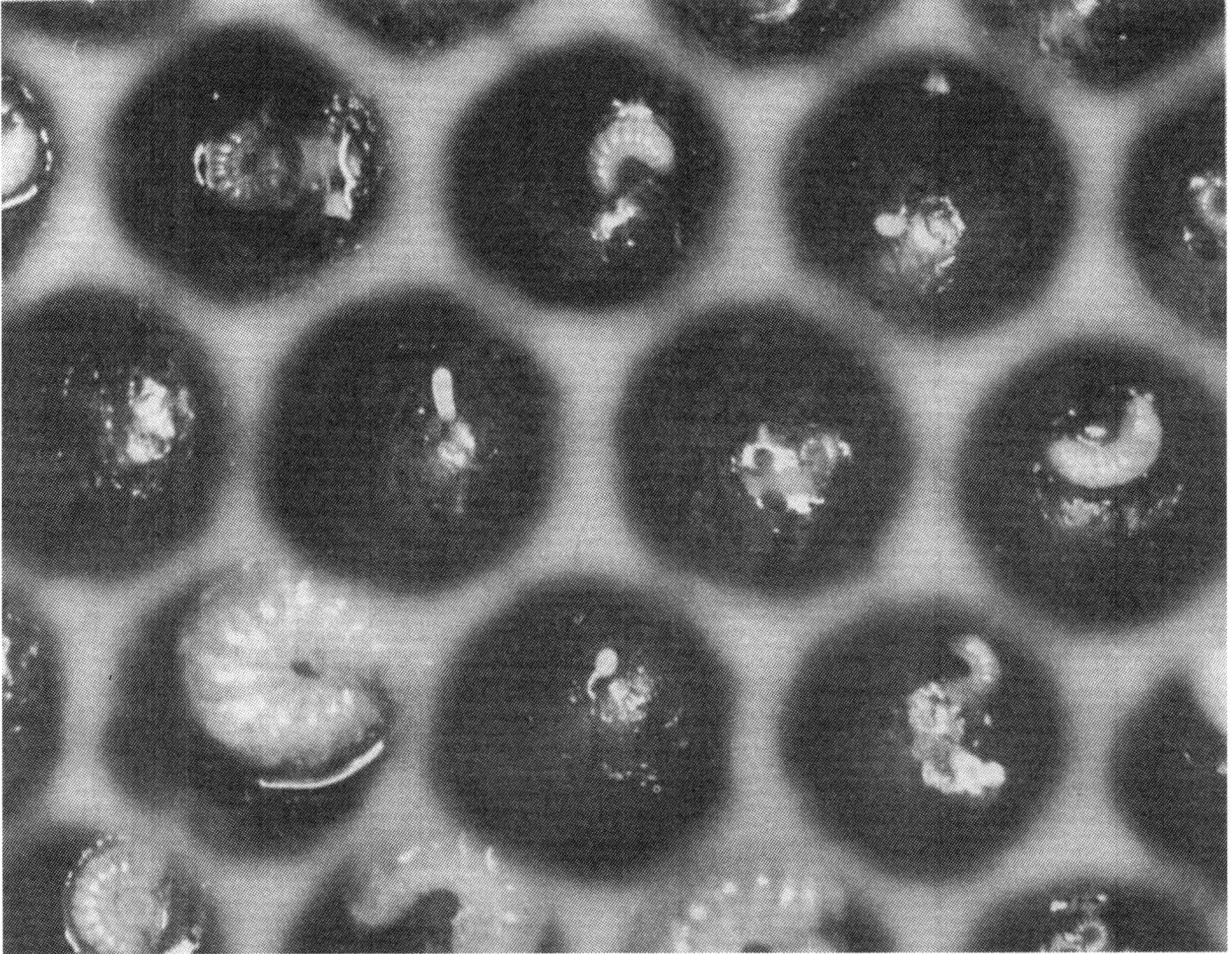

FIGURE 1.—Worker bee brood showing eggs and young larvae.

tine until a larva is nearly mature. In this way, body wastes are stored internally and the food surrounding each larva is protected from fecal contamination. The feces are expelled and pushed down to the bottom of the cell about the time the cocoon is made and after the larva has finished eating.

All castes of honey bees molt about every 24 hours during the first 4 days of larval life. When the ecdysis or molting occurs, the skin splits over the head and slips off the posterior end of the larva. This process normally takes less than 30 minutes. Each new larval stage (instar) is at first only slightly larger than the previous one, but it grows rapidly. The fifth larval instar gains about 40 percent of the total mature larval weight during days 8 and 9 (table 1).

By the end of the 8th day after the egg was laid, the cell containing the worker larva is capped. During the 9th day, the larva spins a cocoon using silk from a special gland in its head. On the 10th day, the larva stretches out on its back with its head toward the cell opening and becomes quiescent inside its cocoon. This stage usually is called the prepupa. The 5th molt, which occurs during the 11th day, reveals the pupal form—white in color and motionless (fig. 2). Color develops gradually, first in the eyes (13th day), then in the abdomen (15th day), legs (16th day), wings (18th day), and finally in the antennae (20th day).

TABLE 1.—*Moults of the honey bee*

Day	Workers		Queens		Drones	
	Stages	Moults	Stages	Moults	Stages	Moults
1						
2	Egg		Egg		Egg	
3		(Hatching)		(Hatching)		(Hatching).
4	1st larval	1st moult	1st larval	1st moult	1st larval	1st moult.
5	2d larval	2d moult	2d larval	2d moult	2d larval	2d moult.
6	3d larval	3d moult	3d larval	3d moult	3d larval	3d moult.
7	4th larval	4th moult	4th larval	4th moult (sealing)	4th larval	4th moult.
8						
9	Gorging	(Sealing)	Gorging		Gorging	
10						
11	Prepupa	5th moult	Prepupa	5th moult		(Sealing).
12					Prepupa	
13			Pupa			
14						5th moult.
15	Pupa					
16			Imago	6th moult (emerging)		
17						
18						
19					Pupa	
20						
21	Imago	6th moult (emerging)				
22						
23					Imago	6th moult.
24						(Emerging).

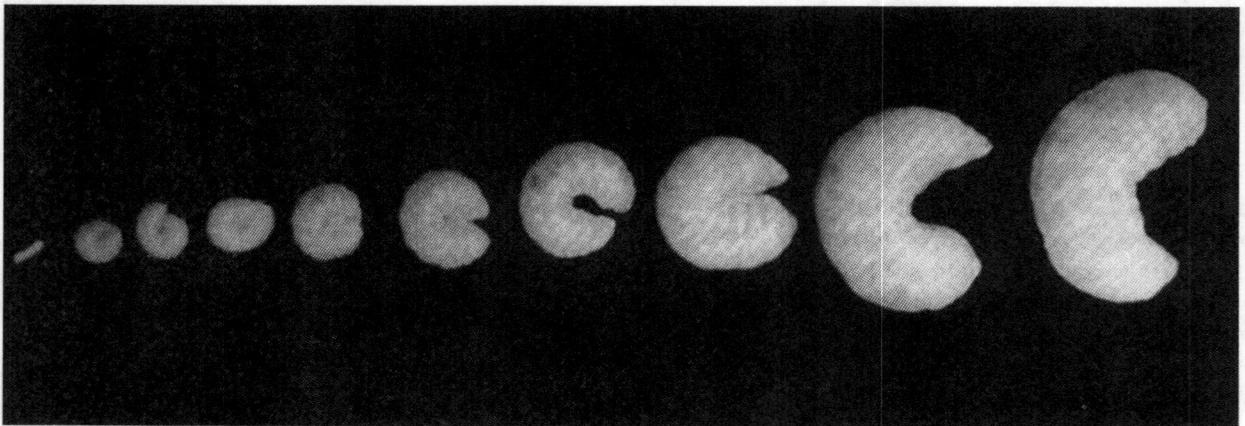

FIGURE 2.—Successive stages of development from egg to mature larva. Enlarged slightly.

Throughout this period, the pupa is encased in a thin outer skin which is shed in the 6th and final molt on the 20th day. Thus, legs, wings, and mouth parts are freed and the pupa becomes an imago (adult) which soon begins to chew its way out of the cell (fig. 3).

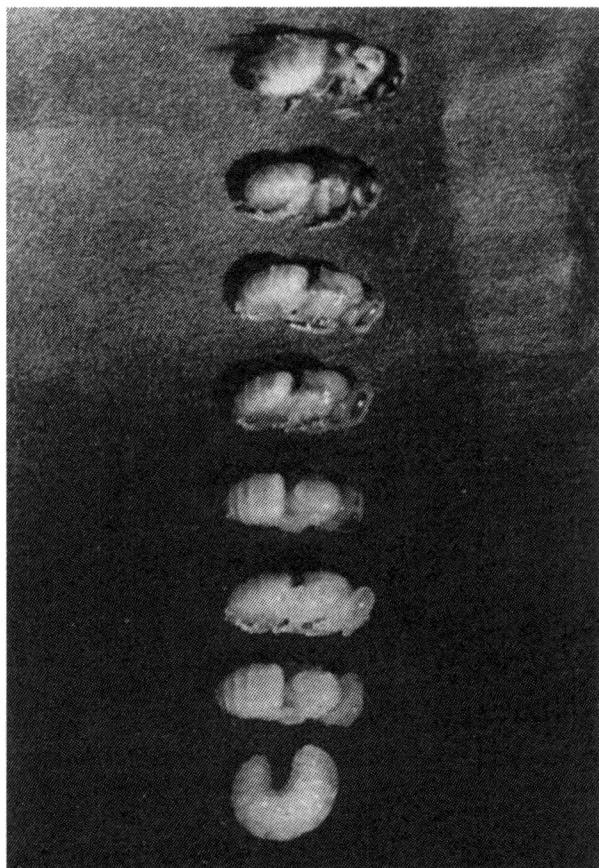

FIGURE 3.—From mature larva to adult bee. About natural size.

Because a bee egg hatches into a larva which becomes first pupa and then imago, bees are said to have a complete metamorphosis. The length of the egg stage (3 days) is the same for all three castes, but the larval and pupal stages are shortest for the queen and longest for the drone (table 1). As with most biological development, the duration may vary between individuals, and the data in table 1 should be considered as close approximations. For example, hatching occurs between 72 and 84 hours, and workers may emerge as early as 19 days or as late as 23 days after the egg was laid.

Activities of Worker Bees

Workers within a honey bee colony engage in various tasks, depending on their age and the needs of the colony. Division of labor by age exists within the worker caste. Bees less than 2 weeks old become involved in cleaning cells and feeding first the older larvae and then larvae of all ages. Workers function as nurse bees during the period that their hypopharyngeal glands are producing brood food.

Older house bees work with honey, pollen, wax, and propolis. Nectar-collecting field bees are met by house bees, usually near the entrance, and are relieved of their nectar loads. The conversion of nectar into honey requires both a physical and a chemical change. The physical change involves the removal of water, which is accomplished by externally manipulating nectar in the mouth parts and then placing small droplets on the upper side of cells and fanning the wings to increase air movement and carry away excess moisture. (Nectar is 30 to 90 percent water, but honey should have no more than 18.5 percent water.) The chemical change requires the addition to nectar of the enzyme invertase, which the bees produce in their salivary glands. This enzyme breaks the disaccharide sugar, sucrose, into two monosaccharide sugars, glucose and fructose. (Nectar may be nearly all sucrose or may contain no sucrose, but honey contains an average of only 1 percent sucrose.)

Pollen pellets are deposited in empty cells near the brood nest by the pollen-collecting workers. In the cells, the pollen undergoes a maturing process to what is commonly called bee bread. Details of the maturing process are not understood.

When bees are about 12 to 15 days old, their wax glands become functional and comb building is possible. Wax scales are removed from between the ventral abdominal sclerites and positioned into place using both feet and mouthparts. Young house bees in the process of comb building hang in festoons and pass the wax scales from bee to bee.

Propolis-collecting bees also serve as propolis storage reservoirs. Propolis is not stored in combs or elsewhere, but is removed from the corbiculae of these field bees and used as needed. House bees fill cracks and cover rough parts with propolis.

During their third week as house bees, workers take short flights for orientation and defecation. Hives painted different colors aid the bees with orientation and reduce the chance of young bees drifting between adjacent colonies. Some of the oldest house bees also serve as guards at the entrance.

After approximately 3 weeks as house bees, the workers become foragers, gathering pollen, nectar, water, and propolis for the colony. This period of their lives also averages about 3 weeks (fig. 4).

Most foragers collect nectar and pollen, but nectar is collected in greater quantities than pollen. Pollen collection tends to be an activity of younger foragers and nectar collection a function of older foragers. Water collectors may comprise 10 percent of all foragers, but this figure is much higher

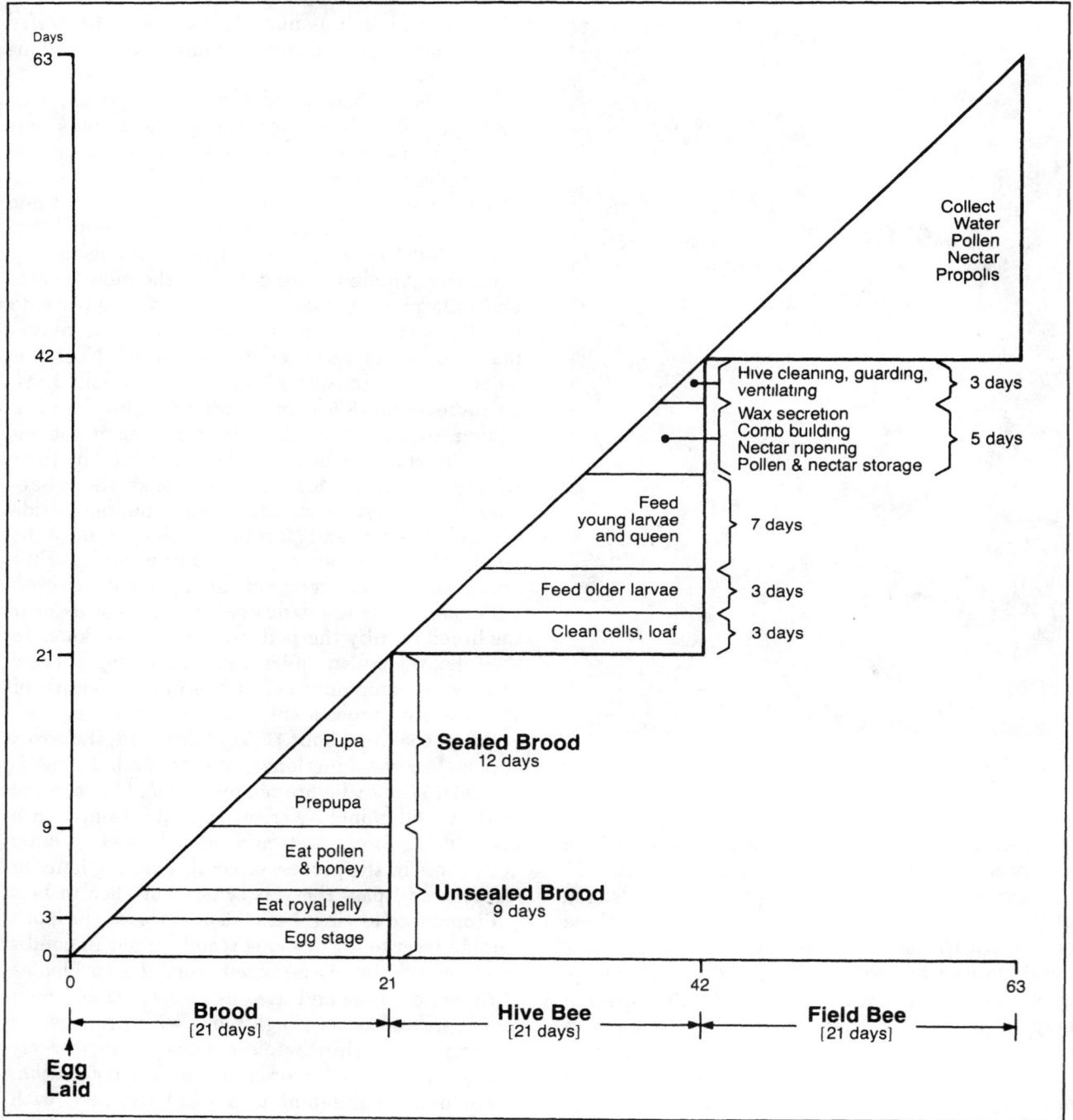

FIGURE 4.—Life history of the honey bee worker.

during periods of heat stress. Propolis collectors are quite rarely observed.

The schedule of worker bee activities is both flexible and reversible, depending more upon physiological age than on chronological age, and is altered according to the needs of the colony. During autumn, a reduction in brood rearing and an increase in pollen consumption result in a population of long-lived "winter" bees having increased fat bodies and protein reserves. The normal 6-week adult life of "summer" bees may be extended to several months in these "winter" bees. Similar extensions of life expectancy also occur when brood rearing is interrupted at other times.

Activities of Drones and Queens

Drones take their first flights at about 8 days of age and are sexually mature at 12 days. Drones fly out on cleansing flights and orientation flights—both of short duration—and also on longer flights to congregation areas in search of a virgin queen. Drones do not forage and spend about three-fourths of their time at complete rest. Their normal lifespan is 8 weeks or less.

Queens newly emerged from their cells are at first ignored but are later touched and licked by workers. This apparently helps prepare the virgin queen physiologically for her mating flight. Mating occurs in drone congregation areas—special locations in the air regularly visited by drones. These occur in the same places year after year.

A queen generally mates 6 to 10 days following emergence. She may go out on several mating flights, mating with several drones on each flight. Additional mating flights are taken until the spermatheca contains an adequate supply of semen

(5 million to 6 million spermatozoa). If mating is delayed more than 3 weeks, there is a high risk of her becoming a drone-layer. Egg-laying usually commences within a week after mating, and a queen can continue to lay fertilized eggs throughout most of her life—usually 2 to 5 years. An old queen will not go out and mate again when her original supply of semen becomes depleted—she simply becomes a drone-layer. An old queen and her supersedure daughter sometimes coexist, thus contradicting the commonly accepted idea of one queen per colony.

References

BERTHOLF, L. M.
　　1925. THE MOULTS OF THE HONEY BEE. Journal of Economic Entomology 18:380.
BUTLER, C. G.
　　1975. THE HONEY BEE COLONY LIFE HISTORY. In The Hive and the Honey Bee, p. 39–74. Dadant & Sons, Hamilton, Ill.
DU PRAW, E. J.
　　1960. RESEARCH ON THE HONEY BEE EGG. Gleanings in Bee Culture 88(2):104–111.
HAYDAK, M. H.
　　1968. NUTRITION DES LARVES D'ABEILLES. In Chauvin, R. Traite de biologie de l'abeille, Vol. I, Biologie et physiologie generales, p. 302–333. Masson Et Cie, Paris.
NELSON, J. A.
　　1915. THE EMBRYOLOGY OF THE HONEY BEE. 282 p. Princeton University Press, Princeton.
────── A. P. STURTEVANT, and B. LINDBURG.
　　1924. GROWTH AND FEEDING OF HONEY BEE LARVAE. U.S. Department of Agriculture Bulletin No. 1222, 37 p.
SNODGRASS, R. E.
　　1975. THE ANATOMY OF THE HONEY BEE. In The Hive and the Honey Bee, p. 75–124. Dadant & Sons, Hamilton, Ill.

SEASONAL CYCLE OF ACTIVITIES IN HONEY BEE COLONIES

By Norbert M. Kauffeld[1]

A colony of honey bees comprises a cluster of several to 60,000 workers (sexually immature females), a queen (a sexually developed female), and, depending on the colony population and season of year, a few to several hundred drones (sexually developed males) (fig. 1). A colony normally has only one queen, whose sole function is egg laying. The bees cluster loosely over several wax combs, the cells of which are used to store honey (carbohydrate food) and pollen (protein food) and to rear young bees to replace old adults.

The activities of a colony vary with the seasons. The period from September to December might be considered the beginning of a new year for a colony of honey bees. The condition of the colony at this time of year greatly affects its prosperity for the next year.

[1] Research entomologist, Science and Education Administration, Carl Hayden Center for Bee Research, Tuscon, Ariz. 85719.

In the fall a reduction in the amounts of nectar and pollen coming into the hive causes reduced brood rearing and diminishing population. Depending on the age and egg-laying condition of the queen, the proportion of old bees in the colony decreases. The young bees survive the winter, while the old ones gradually die. Propolis collected from the buds of trees is used to seal all cracks in the hive and reduce the size of the entrance to keep out cold air.

When nectar in the field becomes scarce, the workers drag the drones out of the hive and do not let them return, causing them to starve to death. Eliminating drones reduces the consumption of winter honey stores. When the temperature drops to 57° F, the bees begin to form a tight cluster. Within this cluster the brood (consisting of eggs, larvae, and pupae) is kept warm—about 93° F—with heat generated by the bees. The egg laying of the queen bee tapers off and may stop completely during October or November, even if pollen is

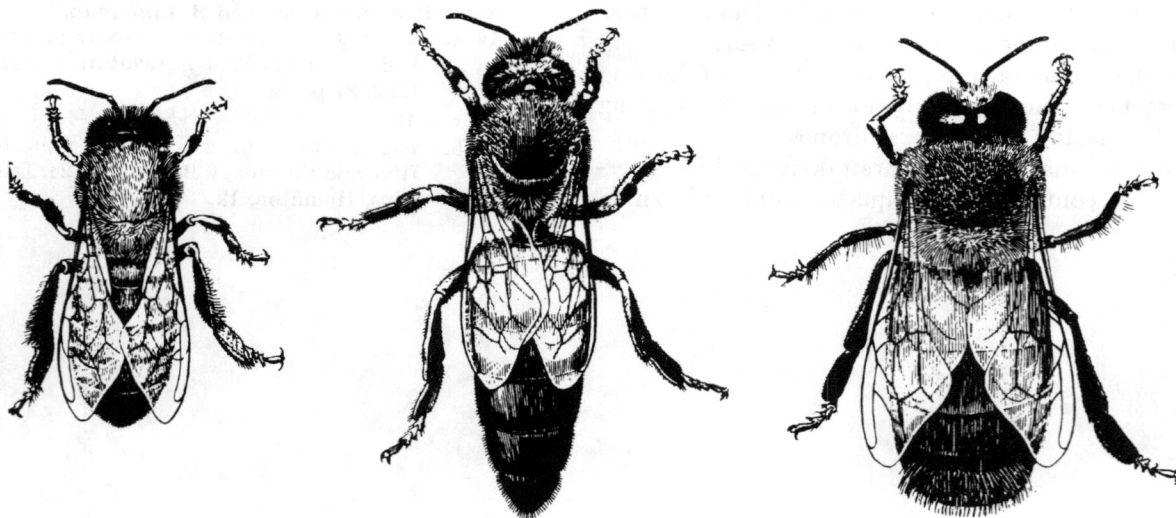

BN-30050

FIGURE 1.—Worker, queen, and drone bees.

stored in the combs. During cold winters, the colony is put to its severest test of endurance. Under subtropical, tropical, and mild winter conditions, egg laying and brood rearing usually never stop.

As temperatures drop, the bees draw closer together to conserve heat. The outer layer of bees is tightly compressed, insulating the bees within the cluster. As the temperature rises and falls, the cluster expands and contracts. The bees within the cluster have access to the food stores. During warm periods, the cluster shifts its position to cover new areas of comb containing honey. An extremely prolonged cold spell can prohibit cluster movement, and the bees may starve to death only inches away from honey.

The queen stays within the cluster and moves with it as it shifts position. Colonies that are well supplied with honey and pollen in the fall will begin to stimulatively feed the queen, and she begins egg laying during late December or early January—even in northern areas of the United States. This new brood aids in replacing the bees that have died during the winter. The extent of early brood rearing is determined by pollen stores gathered during the previous fall. In colonies with a lack of pollen, brood rearing is delayed until fresh pollen is collected from spring flowers, and these colonies usually emerge from winter with reduced populations. The colony population during the winter usually decreases because old bees continue to die; however, colonies with plenty of young bees produced during the fall and an ample supply of pollen and honey for winter usually have a strong population in the spring.

Spring Activity

During early spring, the lengthening days and new sources of pollen and nectar stimulate brood rearing. The bees also gather water to regulate temperature and to liquefy thick or granulated honey in the preparation of brood food. Drones will be absent or scarce at this time of the year.

Later in the spring, the population of the colony expands rapidly and the proportion of young bees increases. As the population increases, the field-worker force also increases. Field bees may collect nectar and pollen in greater amounts than are needed to maintain brood rearing, and surpluses of honey or pollen may accumulate (fig. 2).

As the days lengthen and the temperature continues to increase, the cluster expands further and

FIGURE 2.—A field bee returning to the hive with a load of pollen.

drones are produced. With an increase in brood rearing and the accompanying increase in adult bees, the nest area of the colony becomes crowded. More bees are evident at the entrance of the nest. A telltale sign of overcrowding is to see the bees crawl out and hang in a cluster around the entrance on a warm afternoon.

Combined with crowded conditions, the queen also increases drone egg laying in preparing for the natural division of the colony by swarming. In addition to rearing workers and drones, the bees also prepare to rear a new queen. A few larvae that would normally develop into worker bees are fed a special gland food called royal jelly, their cells are reconstructed to accommodate the larger queen, and her rate of development is speeded up. The number of queen cells produced varies with races and strains of bees as well as individual colonies.

Regardless of its crowded condition, the colony will try to expand by building new combs if food and room are available. These new combs are generally used for the storage of honey, whereas the older combs are used for pollen storage and brood rearing.

Swarming

When the first virgin queen is almost ready to emerge, and before the main nectar flow, the colony will swarm during the warmer hours of the day. The old queen and about half of the bees will rush en masse out the entrance. After flying around in the air for several minutes, they will cluster on the limb of a tree or similar object (fig. 3). This cluster usually remains for an hour or so, depending on the time taken to find a new home by scouting bees. When a location is found, the cluster breaks up and flies to it. On reaching the new location, combs are quickly constructed, brood rearing starts, and nectar and pollen are gathered. Swarming generally occurs in the Central, Southern, and Western States from March to June, although it can occur at almost any time from April to October.

After the swarm departs, the remaining bees in the parent colony continue their field work of collecting nectar, pollen, propolis, and water. They also care for the eggs, larvae, and food, guard the entrance, and build combs. Emerging drones are nurtured so that there will be a male population for mating the virgin queen. When she emerges from her cell, she eats honey, grooms herself for a short time, and then proceeds to look for rival queens within the colony. Mortal combat eliminates all queens except one. When the survivor is about a week old, she flies out to mate with one or more drones in the air. The drones die after mating, but the mated queen returns to the nest as the new queen mother. Nurse bees care for her, whereas prior to mating she was ignored. Within 3 or 4 days the mated queen begins egg laying.

During hot summer days, the colony temperature must be held down to about 93° F. The bees

FIGURE 3.—Hiving a swarm.

do this by gathering water and spreading it on the interior of the nest, thereby causing it to evaporate within the cluster by its exposure to air circulation.

During the early summer, the colony reaches its peak population and concentrates on the collection of nectar and pollen and the storage of honey for the coming winter. After reproduction, all colony activity is geared toward winter survival. Summer is the time for storage of surplus food supplies. The daylight period is then longest, permitting maximum foraging, although rain or drought may reduce flight and the supply of nectar and pollen available in flowers. It is during the summer that stores are accumulated for winter. If enough honey is stored, then the beekeeper can remove a portion and still leave ample for colony survival.

BEE BEHAVIOR

By Stephen Taber III[1]

Bee behavior refers to what bees do—as individuals and as a colony. By studying their behavior, we may learn how to change it to our benefit.

Two practical discoveries of bee behavior made our beekeeping of today possible. One was the discovery by Langstroth of bee space. The other was the discovery by G. M. Doolittle that large numbers of queens could be reared by transferring larvae to artificial queen cups. The discovery of the "language" of bees and of their use of polarized light for navigation has attracted considerable interest all over the world.

Much has been learned about the behavior of insects, including bees, in recent years. As an example, the term "pheromone" had not been coined in 1953, when Ribbands summarized the subject of bee behavior in his book, *The Behaviour and Social Life of Honeybees.* A pheromone is a substance secreted by an animal that causes a specific reaction by another individual of the same species. Now many bee behavior activities can be explained as the effect of various pheromones.

Recently, we have learned how certain bee behavior activities are inherited, and this information gives us a vast new tool to tailor-make the honey bee of our choice. Further studies should reveal other ways to change bees to produce specific strains for specific uses.

The Honey Bee Colony

The physical makeup of a colony has been described. An additional requirement of a colony is a social pattern or organization, probably associated with a "social pheromone." It causes the bees to collect and store food for later use by other individuals. It causes them to maintain temperature control for community survival when individually all would perish. Individuals within the colony communicate with each other but not with bees of

another colony. Certain bees in the colony will sting to repel an intruder, even though the act causes their death. All of these, and perhaps many other organizational activities, probably are caused by pheromones.

There is no known governmental hierarchy giving orders for work to be done, but a definite effect on the colony is observed when the queen disappears. This effect seems to be associated with a complex material produced by the queen that we refer to as "queen substance." There also is evidence that the worker bees from 10 to 15 days old, who have largely completed their nursing and household duties but have not begun to forage, control the "governmental" structure. Just what controls them has not been determined.

These and many other factors make an organized colony out of the many thousands of individuals.

The Domicile

When the swarm emerges from its domicile and settles in a cluster on a tree, certain "scout bees" communicate to it the availability of other domiciles. At least some of these domiciles may have been located by the scout bees before the swarm emerged. The various scouts perform their dances on the cluster to indicate the direction, distance, and desirability of the domiciles. Eventually, the cluster becomes united in its approval of a particular site. Then the swarm moves in a swirling mass of flying bees to it. Agreement always is unanimous.

When a swarm or combless package is placed in a box, allowed to fly, and supplied with abundant food, it builds comb. With a laying queen present, the first comb is "worker" in design, with about 25 cells per square inch. As the population of bees grows larger, and after there is a considerable amount of worker comb built, comb containing larger cells is constructed. This comb, termed storage comb by Langstroth, is used for rearing drones. We have found that bees store their first honey all across the top of the combs, utilizing both drone and worker cells.

[1] Apiculturist, Science and Education Administration, Carl Hayden Center for Bee Research, Tuscon, Ariz. 85719.

The space between honey storage combs is much more uniform than between brood combs. The space left between capped honey cells is usually one-fourth inch or even less—room enough for one layer of bees to move.

As the colony ages, the combs that were first used for rearing worker bees may be converted to honey storage comb; areas damaged in any way are rebuilt. These changes usually affect the bee space and result in combs being joined together with "brace" comb. Strains of bees show genetic variation in building these brace combs.

All these cells are horizontal or nearly so; vertical cells are used for rearing queens. Why horizontal cells are used for the rearing of brood and for honey and pollen storage, whereas vertical cells are built only for queen production, is unknown.

Flight Behavior

When several thousand bees and a queen are placed in new surroundings—which happens when the swarm enters its new domicile or a package of bees is installed, or a colony is moved to a new location—normal flight of some workers from the entrance may occur within minutes. If flowering plants are available, bees may be returning to the hive with pollen within an hour. Bees transferred by air from Hawaii to Louisiana and released at 11:30 a.m. were returning to the new location with pollen loads within an hour. Package bee buyers in the Northern States have noticed similar patterns in bees shipped from the South.

What causes this virtually instant foraging by bees? What determines whether they collect pollen, nectar, or water? If food and water in the hive are sufficient, why should they leave to forage?

Answers to these questions may lead to our directing bees to specific duties we desire accomplished.

Housecleaning

Certain waste material accumulates in a normal colony. Adult bees and immature forms may die. Wax scales, cappings from the cells of emerging bees, particles of pollen, and crystallized bits of honey drop to the floor of the hive. Intruders, such as wax moths, bees from other colonies, and predators, are killed and fall to the floor. Worker bees remove this debris from the hive.

The cleaning behavior of some strains of bees, associated with removal of larvae and pupae that have died of American foulbrood, is genetically controlled by two genes. This discovery is important not only because it might help in developing bees resistant to diseases, but also in indicating that other behavior characteristics of bees can be genetically modified to suit special needs.

Known Pheromone Activity

Chemicals that bees and other insects produce that influence, or direct, behavior of other bees are broadly called pheromones. In honey bees these chemicals are produced by the queen, workers, and probably drones. A list of the known chemicals associated with the queen and worker is given in tables 1 and 2.

This is an interesting and new area for bee research, as this list represents just a beginning. Research has indicated the existence of many other pheromones, which are as yet undocumented. If interested in this topic, consult the technical work listed in Gary (1974).

TABLE 1.—*Queen pheromones*

Gland or source and chemical	Behavior reactions in colonies	Citations [1]
Mandibular:		
9-oxodecenoic acid	Recognition of queen and reduction of egg laying by workers.	Butler (1964).
10-hydroxydecenoic deconic acid		
9-oxodec-trans-2-enoic acid	Mating attractant	Gary (1962).
Do	In combination with worker bees, scent gland holds swarming bees together.	Morse (1971).

[1] Citations are not listed; consult Gary (1974).

TABLE 2.—*Worker bee pheromones*

Producing gland or source	Chemical compound	Behavior reaction in colony	Citations [1]
Nassanoff or scent	Geraniol	Fanning attractant	Boch (*1963*).
Do	Nerolic acid	_____do	Boch (*1964*).
	Geranic acid	_____do	
Do	Citral	_____do	Shearer (*1966*).
Do	All compounds of scent gland.	Swarm attraction and stabilization.	Morse (*1971*).
Sting	Iso-pentyl acetate	Colony alarm	Boch (*1962*).
Mandibular	2-heptanone	Alarm communications	Boch (*1965*).

[1] Citations are not listed; consult Gary (*1974*).

Pheromonal bee behavior activity patterns are easily observable. Nassanoff or scent gland activity is best seen when a swarm is hived. When the bees first enter the new domicile, some bees stand near the entrance and fan. At the same time, they turn the abdominal tip downward to expose a small, wet, white material on top of the end of the abdomen. This seems to affect the other bees, for within several minutes all will have entered the new hive. When bees find a new source of food, they also mark it with the same chemical.

Colony odor refers to the odor of one colony. Because each colony odor is different, colonies cannot be combined into one hive without the bees fighting and killing one another. This odor probably results from a combination of endogenous (pheromone or pheromonelike) materials) and exogenous (food) materials in each hive and seems to be recognizably different for every colony.

When colonies are to be combined, the beekeeper usually places a newspaper between the two sets of bees. By the time the bees have eaten through and disposed of the newspaper, their odors have intermingled and become indistinguishable. During heavy honey flows, differences between colonies seem to disappear, or be submerged by the scent of nectar, and colonies can be united without difficulty.

One of the most interesting and complex pheromones, originally termed "queen substance," is now believed to be a complex of different chemical pheromone compounds which stimulate a large number of complex behavior responses. Its presence in virgin queens in flight attracts the drone for mating from an unknown distance. Its presence in virgin and mated queens prevents the ovaries of the worker bee within the hive from developing and the worker bees from building queen cells. It keeps swarming bees near the queen. Its decrease is a cause of swarm preparation or supersedure. Queen substance is produced in glands in the queen's head. The alarm or sting pheromone also may be a complex of pheromones. When a bee stings, other bees in the immediate vicinity also try to sting in the same place. Smoke blown onto the area seems to neutralize this effect.

Cause of Stinging Bees or Temper

The term "temper" of bees refers to their inclination to sting. Many factors influence the temper of bees, and it is a difficult subject to study. Environment of the hive and manipulation by an individual beekeeper certainly influence temper responses of bees. Temper is probably influenced tremendously by the genetics or inheritance of the bee as well as the environment. The Brazilian or Africanized bee is thought to be more genetically prone to sting than bees in the United States.

Temper of bees commonly has been controlled with smoke. Just why and how smoke affects bees is unknown, even though it has been used by beekeepers worldwide for hundreds of years. Furthermore, instructing beginners and novices exactly when and how to use smoke on bees is almost impossible. It is something that is learned from experience.

The following brief instruction might help beekeepers with limited experience: Smoke the entrance gently enough to force guard bees inside, raise cover, smoke gently. Smoke bees only when they fly up from combs toward hands and face. Move slowly and deliberately. Break propolis seals between hive bodies and frames slowly and evenly.

Don't jar or bump combs and bees. During cold weather, propolis joints snap when pried apart unless care is taken. If combs are kept clean of propolis and burr and brace comb and if care is taken not to crush bees when moving combs and supers, they can be kept quite gentle.

Great care should be exercised in the placement of colonies of bees so that they cannot become a nuisance to friends and neighbors. Bees visiting nearby fishponds, swimming pools, and stock-watering troughs can be a real nuisance as well as dangerous to people and animals. Springtime flight of bees voiding feces and spotting laundry hanging on a line or a new car is irritating. Good public relations are important for beekeepers. Talk to your neighbors about the importance of bees in the community and country at large. Help them to understand that your bees and others are responsible for important pollination and share some honey with them occasionally.

Colony Morale

"Colony morale" generally refers to the well-being of the colony. If the morale is good, the bees are doing what is desired of them, including increasing the colony population, making honey, and pollinating flowers. Many factors affect colony morale. For example, if the queen is removed from a colony during a honey flow, the daily weight gains immediately decrease, although the bee population for the next 3 weeks is unaltered. Also, when a colony is preparing to swarm, the bees practically stop gathering pollen and nectar. Improper manipulations or external environment also affects colony morale. A colony has good morale when the maximum number of bees are making the maximum number of flights to gather nectar and pollen.

Other Methods of Bee Communication

There are other methods of bee communication besides the one involving chemical pheromones. The best known is the "dance" of the returned forager bee so well elucidated by von Frisch and his many students, particularly M. Lindauer.

This dance is so precise that it tells other bees not only in which direction to go but also how far to fly in search of food. This was the first non-human language to be interpreted. The experiments on bee communication by dances were done with dishes of sugar water and not under true foraging conditions of bees collecting nectar from plants. When a returning forager comes back to the hive after finding a highly attractive 100-acre field of sweetclover, does she direct bees to the spot she was working or to the whole field? The last word in dance communication of bees certainly has not yet been written.

Even the most uninitiated are familiar with the soft quiet hum of bees collecting nectar and pollen on their foraging trips. In the hive itself, there are many more bee noises or sounds which are much more subtle. Experienced beekeepers recognize a difference in sound between a colony with a queen and one without. Individual queens and even worker bees emit squeaky sounds called "piping" and "quacking." The bee literature is full of many explanations of the causes and meanings of these sounds. Since these sounds and other hive sounds are now under careful scientific scrutiny, it is really premature to say definitely that they have certain defined meanings. This field of interest may produce useful information in the future.

According to von Frisch, when a bee returns from a foraging trip and dances, she also communicates the kind of "plant" or "flower" on which she was foraging by releasing the perfume of the flower through nectar regurgitation or from nectar aroma on body hairs. Again, most of these experiments were done with dishes of sugar water impregnated with essential oils or plant extracts. These experiments have prompted other experiments that were designed to train bees to work desired crops for pollination. These experiments were unsuccessful. The reason for the failures may well be that the bee language code has not been completely translated. We are still unable to "talk" effectively to the bees and "tell" them what we want done.

Von Frisch also discovered that bees recognize and are guided to flowers by different colors but are unable to communicate these colors. He also showed that the bee's eyes are receptive to polarized light and that polarization of the light from the sky aids the bee's navigation. How light of different wave lengths or intensity affects what goes on inside a hive is being studied.

Age Levels of Bees Correlated With Work Habits

The honey bee is adaptable to many environments. Honey bees that were native only to

Europe, Asia, and Africa have adapted well to all but the polar regions of the world. Part of this adaptability lies in the capacity of the individual bee to "sense" what must be done, then to perform the necessary duty.

Under normal conditions, all ages of bees are in the hive and, in general, the bee's age determines its daily activity. In response to special needs of the colony, however, bees are capable of altering the division of labor according to age. Young bees feed larvae, build comb, and ripen nectar into honey in a rather definite sequence. After about 3 weeks, they become field bees. If many field bees are killed by pesticides, young bees go to the field at a younger age to get necessary chores accomplished.

The Performance of Colonies

Genetically, we found that some bees produce more honey than others, but we do not know why. The individual bee may collect more because of its own genetic inheritance. The colony may store more honey because of the queen's inherited ability to lay more eggs, resulting in a greater total population of bees in the hive, or because the bees are inherently longer lived.

We can affect the bee's environment in conjunction with its inheritance, and our aim is to have good-quality bees and maintain the best colony morale possible. A beekeeper's disturbance of the colony during the honey flow results in a marked decrease in the amount of honey stored for that day and even the following day. Colonies of bees should not be needlessly disturbed; however, some manipulation associated with many aspects of management is necessary.

Bee behavior toward different plants varies greatly. Some plants are particularly attractive for nectar or pollen; others are not. Strains of bees can be genetically selected to visit certain plants, and plants can be selected to be more attractive to bees. Attractive nectar or pollen, or both, can be important in ensuring pollination of bee-pollinated crops. Nectar and pollen availability in plants can be accidentally eliminated by breeding. When this occurs, there is a loss of a potential honey crop, but more important can be the loss of a seed or fruit crop because the plant no longer attracts pollinators. If plants such as soybeans, which cover enormous acreages, could be made more attractive to bees, honey and possibly soybean yields could be greatly increased.

A behavior characteristic of honey bees limits their effectiveness in pollinating some crops. Individual bees usually confine their foraging area in a series of trips to the field to a relatively small area such as a single fruit tree. On the other hand, the foraging area of a colony may comprise several square miles; honey bees flying 2½ miles in all directions from a single hive have access to 12,500 acres. This characteristic and the fact that honey bees distribute themselves well over the area within flight range are important in locating and harvesting available nectar and pollen.

Control of Foraging

A major crop pollination goal is to control foraging bees and get them to more effectively visit and pollinate crops; conversely, we would like to repel them from areas where there is danger from insecticides or where they endanger people. Work with other insects—both social and nonsocial—indicates that this might be accomplished some day by chemical and physical means.

There is considerable evidence that different plant species produce varying attractant compounds associated with their nectar and pollen. Bees are highly attracted to the scent of recently extracted honeycomb and to the scent of honey being extracted or heated. Obviously, chemical scents of certain flowers and to some extent scents incorporated in the collected honey are attractive to bees or associated with available food.

Some pollens also contain chemical compounds that stimulate collection response in bees. Isolation and identification of these bee-attractive compounds and the application of the attractant to plant areas or altering attractants through plant breeding are an area of research of potential importance to crop pollination.

Research should not be confined to chemicals alone, but should be shared equally with various physical factors that can possibly attract or repel bees. In other entomological fields, research on physical methods of controlling insects is receiving intensive investigation. Different insects respond in differing ways; they are attracted to certain light wavelengths and repelled by others. Night-flying moths are repelled or go into defensive maneuvers because of bat sonar signals, whereas crickets and other members of their insect group can be collected by reproducing certain stridulations.

Other Behavior Activities of Bees

The Drones

The time of day that drones fly in search of a mate depends on many factors, such as the geographic location, day length, and temperature. Drones usually fly from the hive in large numbers between 11 a.m. and 4:30 p.m. Morning or early afternoon flights may last 2 or 3 hours. Later flights are shorter. When out of the hive, drones congregate in "mating areas," which may serve to attract virgin queens. These areas usually are less than 100 feet from the ground and seem to be associated with land terrain.

The Queen

The virgin queen becomes sexually mature about 5 days after emergence. She is relatively quiet in the morning and most active in the afternoon. She may begin her mating flights 5 or 6 days after emergence and go on a number of flights over several days. Mating with 8 to 12 drones will stock her spermatheca with 6 million to 7 million sperm. She will begin to lay eggs in 2 to 5 days and may continue for years.

A young, fully mated queen rarely lays drone eggs before she is several months old. After that time, she controls the sex of the offspring by laying either fertilized or nonfertilized eggs.

Worker bees occasionally kill their queen. More frequently, they will kill a newly introduced or virgin queen. To do this, 15 or 20 worker bees collect about her in a tight ball until she starves. Generally, it has been thought that bees "balled" strange or introduced queens because they did not have the proper "colony" odor. The reason for balling is probably more complicated that that, because bees occasionally will ball their own queen. Even if the ball is broken up, the queen seldom survives and the stimulus is powerful enough that the bees taking part in the queen balling are sometimes subsequently balled by other bees.

References

BUTLER, C. G.
    1955. THE WORLD OF THE HONEY BEE. 226 p. Macmillan Co., New York.
VON FRISCH, K.
    1955. THE DANCING BEES. 183 p. Harcourt, Brace & Co., New York.
GARY, N. E.
    1974. PHEROMONES THAT AFFECT THE BEHAVIOR AND PHYSIOLOGY OF HONEY BEES. In Pheromones, M. C. Birch, p. 200–221, North-Holland, Amsterdam, and Elsevier, New York.
HAYDAK, M. H.
    1963. ACTIVITIES OF HONEY BEES. In The Hive and the Honey Bee, 556 p. Dadant & Sons, Hamilton, Ill.
LINDAUER, M.
    1961. COMMUNICATION AMONG SOCIAL BEES. 143 p. Harvard University Press, Cambridge.
RIBBANDS, C. R.
    1953. THE BEHAVIOUR AND SOCIAL LIFE OF HONEY BEES. 318 p. Dover Publications, Inc., New York.

HONEY BEE NUTRITION AND SUPPLEMENTAL FEEDING

By L. N. Standifer [1]

The natural diet of the adult honey bee is pollen and honey. Sometimes, however, when nectar is not available, bees collect sweet-tasting juices from overripe fruit and plant exudates. Also, certain insects secrete honeydew, which bees may collect and store as honey. During periods when no pollen is available, bees may collect powdery animal feed or spores from plants and store this material as they would pollen. This may have some food value but does not sustain brood rearing and is considered a poor substitute for pollen.

Nutritional Requirements

Honey bees require proteins (amino acids), carbohydrates (sugars), lipids (fatty acids, sterols), vitamins, minerals (salts), and water, and these nutrients must be in the diet in a definite qualitative and quantitative ratio for optimum nutrition.

Proteins and Amino Acids

Adult worker bees (1 to 14 days old) obtain dietary protein from pollen which workers collect and bring back to the hive; adult drone bees (1 to 8 days old) obtain dietary protein from food supplied by young workers which is a mixture of glandular secretions, pollen, and honey; and larval and adult queens obtain their protein from royal jelly secreted by young worker bees. Royal jelly also is fed to worker larvae less than 3 days old. Royal jelly is a secretion of the hypopharyngeal glands of worker bees normally 5 to 15 days of age. It is a creamy, milky white, strongly acid substance with a moisture content of 65 to 67 percent and rich in protein lipids, reducing sugars, B vitamins, vitamin C, and minerals.

Honey bees require specific amino acids for normal growth and development, reproduction, and brood rearing. The protein and amino acid requirements of larval and adult queens are un-

[1] Entomologist, Science and Education Administration, Carl Hayden Center for Bee Research, Tuscon, Ariz. 85719.

known, but we have a fairly comprehensive knowledge of the chemical constitution of their basic food, royal jelly.

During the first 5 or 6 days of adult life, worker bees consume large amounts of pollen to obtain the protein and amino acids required to complete their growth and development. If young adult worker bees do not consume needed proteins, their hypopharyngeal glands (brood food glands) will not develop completely, and their royal jelly will not support normal growth and development of worker larvae or egg production in the adult queen. The requirement for protein decreases when worker bees discontinue nursing (between 10th to 14th day of adult life). Subsequently, the chief dietary constituent becomes carbohydrates obtained from nectars and honey.

Carbohydrates

Carbohydrates are abundant in the natural diet of the honey bee and are used mainly for the production of energy, but may be converted to body fats and stored. Some carbohydrates can be utilized by bees, some cannot, and some are toxic. Adult bees thrive on glucose, fructose, sucrose, trehalose, maltose, and melezitose, but they cannot use rhaminose, xylose, arabinose, galactose, mannose, lactose, raffinose, dextrin, or insulin. Differences in carbohydrate utilization between larvae and adults may be due to the absence of appropriate enzymes.

Lipids

Information on the nutritional need for dietary lipids (fatty acids, sterols, and phospholipids) in honey bees is fragmentary and inconclusive. Generally, lipids are used for energy, synthesis of reserve fat and glycogen, and for the functioning of cellular membranes. The lipid composition of adult bees differs from that of pollen. However, a phospholipid found in pollen also is found in the body tissue of adult bees. Another substance, 24-methylene cholesterol, also found in pollen, is the major sterol of the body tissue of adult queen and worker bees. Possibly, certain lipids have a significant role in the lubrication of food when it is ingested and prepared for absorption. All insects studied critically were found to require a dietary sterol; therefore, it is reasonable to assume the honey bee also requires this lipid.

Vitamins

When bees begin producing royal jelly for the young larvae and the queen, they need a diet high in vitamins. Nurse bees seem to need the following vitamin B complex for brood rearing: thiamine, riboflavin, nicotinamide (niacin, nicotinic acid), pyridoxine, pantothenate (pantothenic acid), folic acid, and biotin. Pantothenic acid is needed in worker-queen differentiation and nicotinic acid, in initiating brood rearing. In addition to these vitamins, ascorbic acid (vitamin C) also seems essential for brood rearing.

In general, the vitamin needs of a honey bee colony are satisfied as long as the pollen stores are abundant in the hive or fresh pollen is available to bees in the field. Micro-organisms naturally present in the alimeatry canal of bees may provide vitamins, and other essential substances, which may make an otherwise unsuitable diet adequate.

Minerals

Minerals required in the diet of humans and other vertebrates (sodium, potassium, calcium, magnesium, chlorine, phosphorus, iron, copper, iodine, manganese, cobalt, zinc, and nickel) are needed by some species of insects. Pollens contain all these minerals, some of which are required by bees.

Water

Water is collected by bees and used primarily as diluent for thick honey, to maintain optimum humidity within the hive, and to maintain appropriate temperatures in the brood area. The amount of water required and collected by a colony is generally correlated with the outside air temperature and relative humidity, strength of colony, and amount of brood rearing in progress.

Ingestion and Digestion

Food enters the alimentary canal (fig. 1) by way of the month and passes through the *esophagus* to the *honey stomach*. In the honey stomach hydro-

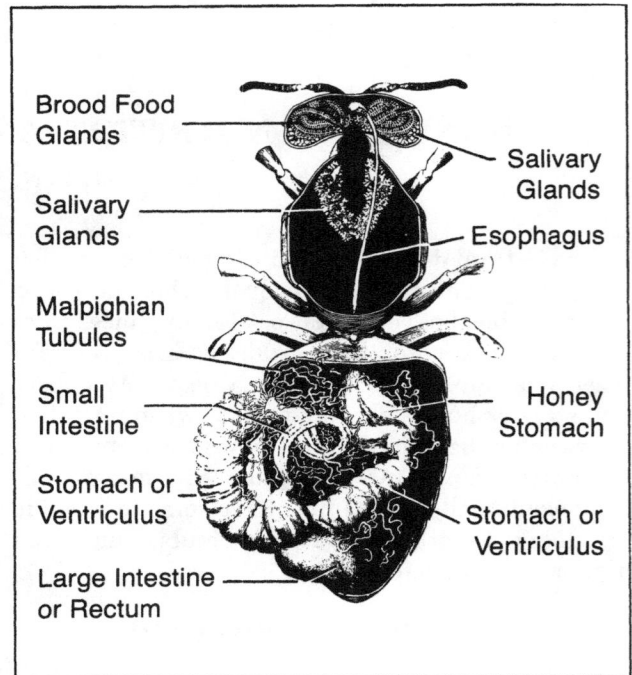

BN-30048

FIGURE 1.—Alimentary canal of adult worker honey bee.

lyzing enzymes break down the principal sucrose of nectar to the simpler monosaccharides glucose and fructose present in honey. Immediately behind the honey stomach is the *proventricular value* or honey stopper. It retains the nectar load in the honey stomach, controls passage of food into the *midgut* or *ventriculus*, and prevents food substances in the midgut from returning to the honey stomach.

The midgut is a relatively large segment of the alimentary canal, where food is temporarily stored and most digestion occurs. The inner wall of the midgut is lined with a *peritrophicm embrane*, presumably to protect the cells from damage by the gut contents.

The alimentary canal is completed by a short *small intestine* and a large *intestine* or *rectum* that comprises the *hindgut* where food digestion is completed. Undigested food residues are reformed into feces in the rectum and eliminated through the terminal *anus*. Passage of pollen through the alimentary canal of adult bees requires about 2½ hours. Feces of adult bees contain almost intact, empty pollen grain shells.

The complex foods ingested by bees must be broken down (digested) into simpler units before they pass through (absorbed) the gut wall into the hemolymph (blood) for ultimate assimilation and utilization. Digestion depends on the activity of enzymes. Enzymes are present in the secretions of the salivary, postcerebral, and hypopharyngeal glands and in the secretions of the midgut epithelial cells. In addition, digestion may be facilitated by the micro-organisms present in the alimentary canal. Compound sugars must be broken down by enzymes to simple sugars before they can be absorbed and utilized.

Bees apparently do not have the enzymes or micro-organisms needed to digest the complex carbohydrates (cellulose, hemicellulose, and pectin) in the outer wall of pollen grains. Enzymes gain access to food inside punctured pollen grains and also by dissolving the "soft germinal pore areas" with digestive enzymes. Enzymes that digest protein are abundant in the alimentary canal of the adult bee and are furnished almost entirely by the midgut and hypopharyngeal glands. Proteins are first broken down to peptones and polypeptides; and these, in turn, are hydrolyzed to amino acids.

The lipid-splitting enzyme lipase is abundant in the midgut of adult workers and drones. In higher animals, lipids are digested by lipase or esterases into free fatty acids and glycerol. The fatty acids are made water soluble by neutralization with alkalies in the alimentary canal. Some insects produce enzymes that hydrolyze certain phospholipids (that is, ecithin and spingomyelin), but probably digestion of the esters and fatty acids usually results from the activity of bacteria. Certain lipids may be absorbed unchanged also.

Food absorption begins in the upper portion of the large intestine and is completed in the rectum, where water salts and other organic molecules are selectively absorbed. There are two pairs of rectal glands or pads on the sides of the rectum that function in water and possibly fat absorbtion.

Sources and Chemical Composition of the Natural Foods

Nectar

Nectar is the major source of carbohydrate in the natural diet of honey bees. It may contain 5 to 75 percent soluble solids (sugars) although most nectars are in the 25- to 40-percent range. The primary sugars are sucrose, glucose, and fructose. As nectar is manipulated and finally stored as honey, much of the sucrose is inverted to approximately equal parts of glucose and fructose. A normal-sized honey bee colony may use the nectar equivalent of 300 to 500 pounds a year.

Pollen

Pollen is eaten by adult bees and fed (via the mixture of glandular secretions, honey, and pollen supplied by nurse bees) to worker and drone larvae after they are 3 days old. Unlike young house bees, field bees do not require pollen in their diet. Stored pollen (bee bread) is consumed by nurse bees (fig. 2). Under natural conditions, pollen collected by bees is usually stored on the periphery of the brood area (fig. 3). In a colony rearing brood, pollen placed next to a comb full of eggs is consumed in 2 or 3 days; if placed on the periphery of larvae, it is used by nurse bees within 1 or 2 days but it may be stored for much longer periods of time. A normal-sized colony may consume 100 pounds or more pollen a year. Not all pollens are nutritionally alike; bees generally collect and utilize a mixture, and many individual pollens are nutritionally inadequate.

The protein content of pollens varies from 10 to 36 percent. Some pollens contain proteins that are deficient in certain amino acids required by bees. All the amino acids listed in table 1, except threonine, are essential for normal growth of the

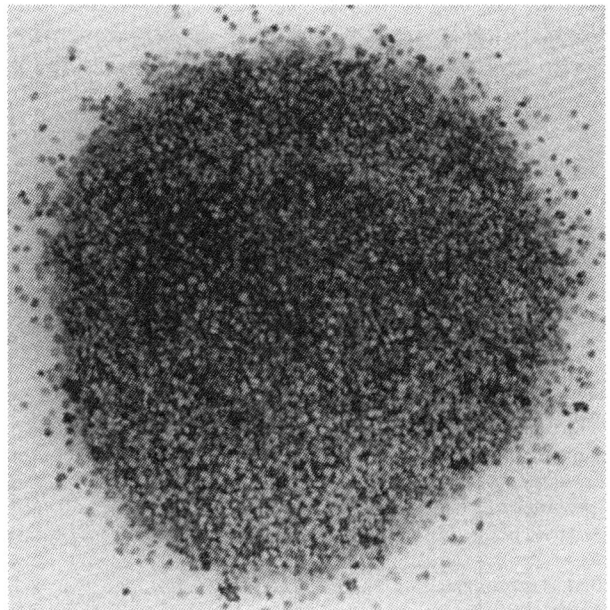

FIGURE 2.—Pollen pellets trapped from bees on their return to the hive.

FIGURE 3.—Two types of pollen trap: *Left*, Emmett Harp type; *right*, O.A.C. type.

young adult bee. With the exception of histidine and perhaps arginine, they cannot be synthesized by bees and must be obtained from the consumed pollens or from some other appropriate protein source.

An average pollen mixture contains lipids (fats) and the following minerals: calcium, chlorine, copper, iron, magnesium, phosphorus, potassium, silicon, and sulfur. Vitamins include ascorbic acid, biotin, vitamins D and E, folic acid, mositol, nicotinic acid, pantothenic acid, pyridoxine, riboflavin, and thiamine. Amino acid content is listed in table 1.

TABLE 1.—*Amino acid content of average pollen expressed as percent of crude protein*

Constituent	Average pollen (crude protein, 26.3%)
	Percent
Arginine	5.3
Histidine	2.5
Isoleucine	5.1
Leucine	7.1
Lysine	6.4
Methionine	1.9
Phenylalamine	4.1
Threomine	4.1
Trypotophane	1.4
Valine	5.8

Supplementary Feeding of Honey Bee Colonies

Honey bees can be fed various foodstuffs to supplement inadequate supplies of pollen or honey. In early spring before pollen and nectar are available or at other times of the year when these materials are in short supply, supplementary feeding may help the colony survive or make it more populous and productive. As modern land-use practices reduce dependable nectar and pollen supplies, the need for supplemental food becomes more and more urgent. Colonies provided with adequate stores in the autumn may not need supplemental foods. However, if the spring weather is unusually cold and rainy, colonies may need supplemental foods for subsistence and continued brood rearing until nectar and pollen can be collected. A sudden curtailment of food when broodrearing activities are in progress will result in reduced bee population.

In practice, beekeepers feed their bees supplemental foods to develop and maintain colonies with optimum populations for: (1) nectar flows, (2) pollination of crops, (3) autumn and spring divisions, (4) queen and package-bee production, and (5) overwintering. Supplemental feeding may also be of value for building up colonies after pesticide damage.

Protein Supplemental Foods for Bees

Numerous kinds of plant and animal products have been fed to bees in attempts to find a substitute to replace pollen in their natural diet.

None has been found that is a complete replacement for natural pollen. Certain protein foodstuffs, however, will improve nutrition and ensure continued colony development in places and times of shortage of natural pollen.

Wheast,[1] soybean flour, and several brewer's yeast products—fed singly or in combination—are palatable to bees and contain the quality and quantity of proteins and amino acids, lipids, vitamins, and minerals required for growth and development of individuals and reproduction of the colony. The yeast products and soybean flour formulations presented in this publication can be fed as a dry mix or moist cake inside the hive or as a dry mix in open feeders outside the hive. Bees are unable to collect wheast in its original dry state because of its large particle size; therefore, it must be fed as moist cake inside the hive.

The choice of sugar to use in protein diets depends partly upon the cost of sugar. Equally important is to use a sugar or sugar syrup that will not cause the moist cake to become hard in a few days when exposed to the warm and comparatively dry environment of the brood nest. Sucrose (cane or beet sugar), isomerized corn syrup,[2] and type-50 sugar syrup[3] with protein supplements produce cakes that maintain their consistency for long periods. Cakes prepared with honey maintain their consistency for a prolonged period. However, it is comparatively expensive, and may transmit bee diseases.

The addition of 10 to 12 percent pollen to a supplement fed to bees improves palatability. The addition of 25 to 30 percent pollen improves the quality and quantity of essential nutrients that are required by bees for vital activity. A bulletin providing several formulae for protein supplemental diets has been published by the Department.

To prepare moist pollen supplement cakes for feeding inside the hive, first dissolve the granulated sugar in the volume of water indicated in the formula. Subsequently, add the brewer's yeast,

torula yeast, wheast, soybean flour, or other suggested material to the sucrose syrup and stir thoroughly. Bee-collected pollen pellets should first be dissolved in water (one-third gallon of water for each pound of pollen pellets) since they do not readily soften in sucrose syrup. The pollen-water suspension is then mixed with the sucrose syrup. More water may be necessary in formulae containing pollen. Each beekeeper should experiment with the formulations to determine the amount of water necessary. In humid areas, the suggested amount of water may be excessive.

Do not use pollen from diseased colonies or from an unknown or questionable source. Preferably, the beekeeper should get the pollen from his own disease-free colonies to avoid possible infection from pollen. The pollen should be stored in a freezer or dried and stored in air-tight containers. Dried pollen more than 2 years old loses much of its nutritional value. Pollen can be used to replace other protein material in any formula.

After thoroughly mixing the combined materials, the final product should be of a doughlike consistency. It should then be divided into 1½-pound cakes and wrapped in waxed paper to prevent the loss of moisture. If cakes are not fed on the day prepared, they can be held in a cool place for several days or in a freezer for several weeks without deterioration or loss of food value.

Protein supplements can be fed any time inside the hive on the top bars or in division-board feeders. When weather permits bees to fly, the materials can be fed in the open, in trays, boxes, tubes, or other open containers. When the supplement is supplied to bees as a moist cake inside the hive, it should be in close proximity to the unsealed larvae in brood combs where nurse bees have ready access to it. A dry mix of moist formulae can be prepared by replacing the water with an equivalent amount of sucrose. Dry mix is usually fed in open feeders. To protect it from rain or dew, the feed should be placed under a roof or hive cover where it is readily accessible to the bees (fig. 4).

Therapeutic drugs may be incorporated in the protein or carbohydrate foods for control of bee diseases. Drugs, however, should never be fed to a colony during or 5 weeks before a major nectar flow. All medicants that are presently recommended for the prevention or treatment of brood diseases and Nosema can be fed in pollen and carbohydrate (dry sugar or syrup) supplements.

---

[1] A dairy yeast (*Saccharomyces fragilis*) grown in cottage cheese whey containing 54 to 60 percent protein.

[2] Produced by means of a multiple enzyme-conversion of corn syrup and composed largely of the simple sugars glucose (dextrose) and fructose (levulose).

[3] Contains 77 percent solids by volume, equal parts of dextrose and levulose, 1.1 kilograms of dry weight/liter, 0.5 percent ash, specific gravity of 1.39, and a pH of 4.8 to 5.7. Coe Sales Company, Phoenix, Ariz.

FIGURE 4.—*A*, Feeding pollen supplement in waxed paper, water in sponge in plastic bag, and sugar syrup in a frame feeder inside the hive; *B*, feeding dry pollen substitute outside the hive; *C*, feeding sugar syrup from an inverted pail; *D*, water supply tank with attached metal feeding pan (excelsior on wooden slats prevents drowning and steel grid excludes wildlife).

Carbohydrate Supplemental Foods for Bees

Proper colony management should ensure adequate honey reserves or stores in the hive at all times, but feeding sugar may sometimes be necessary. Whenever the honey supply in the colony is low and nectar in the field is in short supply, or inaccessible due to adverse weather, the colonies should be fed sugar supplement. Brood rearing requires a large amount of honey and pollen.

Cane or beet sugar, isomerized corn syrup, and type-50 sugar syrup are satisfactory substitutes for honey in the natural diet of honey bees. The last two are supplied only as a liquid to bees.

Preparation and feeding

Cane or Beet Sugar Syrup.—For spring feeding, mix one part by volume (two parts for autumn

feeding) of sugar with one part water heated to 50° to 65°C (about 140°F).

Isomerized corn syrup or type-50 sugar syrup.—Dilute syrup with an equal volume of water.

Sugar syrup can be supplied to bees inside the hive by one of the following methods:

1. *Friction-top pail.*—Puncture several holes in the cover and invert pails on the top bars of the frames or over the hole in the inner cover only and place an empty hive and the hive cover over all.

2. *Combs within the brood chamber.*—When some of the combs are empty, sugar syrup can be poured directly into the cells with a sprinkling can.

3. *Division board feeder.*—A container that replaces a comb in the brood nest. A plastic bag type is also available.

4. *Boardman feeder.*—This feeder is widely used, especially by hobby beekeepers. Its capacity is

small, however, and the syrup, being outside, tends to cool down excessively at night. In addition, the air in the feeder may expand during the day, forcing the food out of the container faster than the bees can consume it, and its exposure to other bees may stimulate robbing.

When package bee colonies are established on empty combs or comb foundation, they should be fed thick syrup (two parts sugar and one part water) for 2 or 3 weeks. Colonies also are fed sugar syrup to stimulate brood rearing for queen and package-bee production to meet early shipping date schedules by producers.

Precautionary measures if robbing is a danger

1. Feed late in the day.
2. Disturb the bees as little as possible.
3. Reduce hive entrances.

Feeding dry sugar

As an emergency measure in late winter when it is too early to feed sugar syrup, bees may be fed dry sugar by placing a pound or two on the inverted inner cover. Some beekeepers increase the feeding space by providing a wooden rim on top of the inner cover.

Preparation and feeding of sugar candy

Mix one part sugar with one part water by weight and heat this mixture until it becomes the thickness of fudge (soft-ball stage). Pour the candy mix on waxed paper and allow to harden. Feed by placing the candy on the top bars directly over the brood nest and cluster.

Supplying Bees With Water

A supply of water must be available to bees at all times. A lack of it adversely affects the nutrition, physiology, brood rearing, and normal behavior. If a natural source is not within a half-mile or less, a supply should be provided. Pan or trays in which floating supports—such as wood chips, cork, or plastic sponge—are present may be satisfactory.

The beekeeper who sees that his colonies always have adequate provisions of food and water is likely to have strong, productive colonies.

References

BARKER, R. J.
1977. SOME CARBOHYDRATES FOUND IN POLLEN AND POLLEN SUBSTITUTES ARE TOXIC TO HONEY BEES. Journal of Nutrition 107:1859–1862.

DETROY, B. F., and E. R. HARP.
1976. TRAPPING POLLEN FROM HONEY BEE COLONIES. U.S. Department of Agriculture Production Research Report No. 163, 11 p.

DIETZ, A.
1975. NUTRITION OF THE ADULT HONEY BEE. *In* The Hive and the Honey Bee, chapter V. Dadant & Sons, Hamilton, Ill.

DOULL, K. M.
1977. TUCSON POLLEN SUPPLEMENTS. American Bee Journal 117:266–297.

HAYDAK, M. H.
1970. HONEY BEE NUTRITION. Annual Review of Entomology 15:143–156.

JOHANSSON, T. S. K., and M. P. JOHANSSON.
1976 and 1977. FEEDING SUGAR TO BEES. Bee World 57(4):137–142; 58(1):11–18; 58(2):49–52.

——— and M. P. JOHANSSON.
1977. FEEDING HONEY BEES POLLEN AND POLLEN SUBSTITUTES. Bee World 58(3):105–118.

STANDIFER, L. N.
1967. A COMPARISON OF THE PROTEIN QUALITY OF POLLENS FOR GROWTH-STIMULATION OF THE HYPOPHARYNGEAL GLANDS AND LONGEVITY OF HONEY BEES, "APIS MELLIFERA L." (HYMENOPTERA:APIDAE). Insectes Sociaux. XIV(4):415–426.

——— W. F. MCCAUGHEY, F. E. TODD, and A. R. KEMMERER.
1960. RELATIVE AVAILABILITY OF VARIOUS PROTEINS TO THE HONEY BEE. Annals of the Entomological Society of America 53(5):618–625.

——— and J. P. MILLS.
1977. THE EFFECTS OF WORKER HONEY BEE DIET AND AGE ON THE VITAMIN CONTENT OF LARVAL FOOD. Annals of Entomological Society of America 70(5):691–694.

——— F. E. MOELLER, N. M. KAUFFELD, and others.
1978. SUPPLEMENTAL FEEDING OF HONEY BEE COLONIES. U. S. Department of Agriculture, Agriculture Information Bulletin 413, 9 p.

TYPES OF HIVES AND HIVE EQUIPMENT

By B. F. Detroy[1]

Honey bees existed on earth long before the appearance of humans. In these early times, bees lived in naturally protected sites such as caves, trees, overhangs, and other similar locations. When humans appeared and advanced in knowledge sufficiently to realize the importance of the honey bee as a source of honey and wax, they attempted to control its existence and distribution by providing a suitable abode or "hive." The hive may be defined as a manufactured home for bees.

Early hives were crude shelters made from any suitable material available to man in his particular locale. Probably the first hives were horizontal sections, either rectangular or circular, made of bark with the ends closed by wooden plugs. A perforation in one end provided an entrance for the bees. Other materials used included clay, cork, wood, and straw. Generally, these materials were shaped into cylindrical or basket-shaped vessels in which a small colony of bees could be kept. Some were furnished with wooden bars on which the combs were built, making it easier to remove the honey and wax. Later, this same type of bar was used in various types and sizes of hives made of wood that sometimes could be enlarged.

Hives of this type were used throughout the bee-populated parts of the world until the 16th century, when a series of events advanced the science of beekeeping and resulted in a corresponding advance in hive development. Events mainly responsible for this advance included scientific developments enabling beekeepers to better understand the life cycle and biology of bees, management techniques that provided greater colony control, and the spread of the honey bee to two new continents. This advance in apicultural knowledge led to the eventual discovery of the "bee space" by L. L. Langstroth and his subsequent development of the movable-frame hive in 1851.

---

[1] Agricultural engineer, Science and Education Administration, Bee Management and Entomology Research, University of Wisconsin, Madison, Wis. 53706.

Langstroth patented his hive, and during the next half century many others developed hives using his principle, often violating his patent. Claims of the perfect hive were frequent and often disputed. Differences of opinion centered mainly on the size of the hive, a single hive body being considered enough to satisfy the brood space requirement for an entire colony. During this period, the development of the excluder, extractor, and comb foundation advanced the production of extracted honey and increased the demand for large colonies. The resulting change to production of extracted honey led to the use of multiple-story brood nests for each colony. By the early 20th century, discussion and experimentation led to increased use of the Langstroth hive or the Dadant hive.

The dimensions of the Langstroth hive and frame are given in figure 1. The original Langstroth hive contained 10 frames spaced 1⅜ inches center to center. The Dadant hive contained 10 frames of the Quinby size, 18½ by 11¼ inches, which gave it about the same capacity as a Langstroth hive with 12 frames. The Dadant hive was changed about 1920 to the modified Dadant hive and it is still in existence today. The modified Dadant hive had 11 frames of the Langstroth length and Quinby depth spaced 1½ inches center to center.

Standard Hive

The 10-frame Langstroth hive today in American beekeeping is known as the "standard" hive and is increasing in popularity and usage. Several variations of the standard hive are used to a limited degree and include the 8-frame Langstroth, modified Dadant, 12-frame Langstroth, and shallow square. The shallow square is a square hive that holds 13 frames but is only 6⅝ inches deep. Hives used today are basically of the same design and vary in the size of the hive body and the depth of the frames it contains. The design, dictated largely by the requirements of the bees, is of

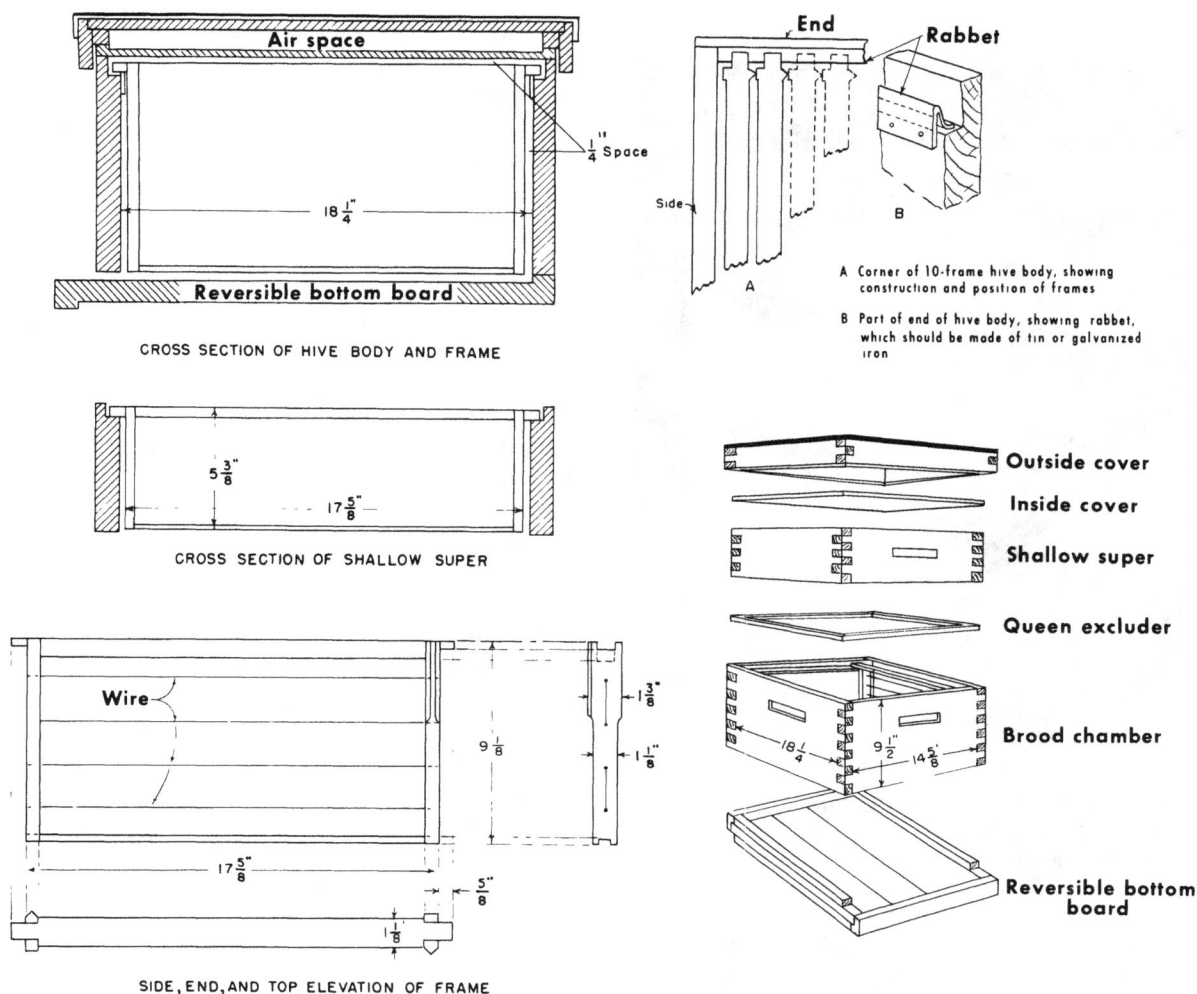

Air space

Reversible bottom board

CROSS SECTION OF HIVE BODY AND FRAME

$18\frac{1}{4}"$

$\frac{1}{4}"$ Space

End

Rabbet

Side

A

B

A Corner of 10-frame hive body, showing construction and position of frames

B Part of end of hive body, showing rabbet, which should be made of tin or galvanized iron

CROSS SECTION OF SHALLOW SUPER

$5\frac{3}{8}"$

$17\frac{5}{8}"$

Outside cover

Inside cover

Shallow super

Queen excluder

Brood chamber

Reversible bottom board

Wire

$9\frac{1}{8}"$

$1\frac{3}{8}"$

$1\frac{1}{8}"$

$17\frac{5}{8}"$

$\frac{5}{8}"$

$1\frac{1}{8}"$

$18\frac{1}{4}"$

$9\frac{1}{8}"$

$14\frac{5}{8}"$

SIDE, END, AND TOP ELEVATION OF FRAME

FIGURE 1.—Plans and dimensions for Langstroth 10-frame beehive.

simple construction and can be used in multiple parts to serve any need.

The modern American beehive consists of a bottom board, usually two hive bodies with frames of drawn comb for the brood nest, supers for the honey crop, inner cover, and outer cover (fig. 2). It is designed simply, lightweight, mobile, economical, and provides the beekeeper maximum opportunity for colony control. The hive bodies that are occupied by the queen and her brood are referred to as brood chambers, and the ones used for storage of surplus honey are called supers. Brood chambers and supers of a hive may be separated by a queen excluder that confines the queen and drones to the brood nest but allows the worker bees to pass into the supers to store surplus honey.

The usual recommendation for housing a single-queen colony in honey-producing areas is three standard brood chambers and three supers, totaling six standard hive bodies. In $6\frac{5}{8}$-inch-deep equipment, four bodies commonly are used for brood chambers and four for supers, totaling eight bodies.

Generally, having all hive equipment of the same dimensions is advantageous to the beekeeper. Brood chambers and supers can then be used interchangeably within or between hives. The "so-called" shallow super may be any one of three depths. Shallow supers used for the production of section comb honey are $4\frac{3}{4}$ inches deep, and those for extracted honey may be either $5\frac{11}{16}$ or $6\frac{5}{8}$ inches deep. All three depths are used for the production of bulk comb honey. Some beekeepers use $6\frac{5}{8}$-inch hive bodies for brood chambers.

FIGURE 2.—Modern beehive cut away to show interior and placement of movable frames: *Bottom*, full-depth hive body; *middle and top*, shallow hive bodies.

Bottom boards usually are reversible, so that a full-width bottom entrance of either ⅞-inch or ⅜-inch depth may be used. Either of these entrances may be restricted to any desired opening size by use of entrance blocks or cleats. The reversible bottom board can be easily removed for cleaning and may be used with or without a hive stand. Covers are used to protect the top of the hive and may be either of two basic types, the flat cover or the telescoping cover. The flat cover is made of a flat board with cleats on opposite ends extending down over the front and back of the hive and is used without an inner cover. The telescoping cover extends over the top of the hive to a depth of an inch or more, is covered with sheet metal, and is generally used with an inner cover. Inner covers contain an oblong hole to accommodate a bee escape and to provide an opening for feeding sugar syrup.

Hoffman Frame

The "Hoffman"-style frame is used almost exclusively in America. This type of frame is self-spacing through contact of the wider top portion of the end bars of adjacent frames. The frame is designed to receive foundation, and the end bars contain holes through which wires may be installed for embedding in the foundation to provide support. Foundation is a sheet of pure beeswax embossed to correspond to the bases of honeycomb cells. Foundation is available in various types and thicknesses. Thin foundation generally is used for comb honey production, and crimp-wired foundation commonly is used for brood and extracting combs.

Aluminum and plastic foundations have been developed and consist of a thin sheet of the material embossed with the cell base and coated on both sides with a thin layer of beeswax. Neither of these foundations is readily accepted by bees unless properly installed in frames and given to the bees during a heavy honey flow. An aluminum foundation is not suitable in the brood nest of the hive, probably because aluminum is a good conductor of heat.

Recently, hive equipment has been made of materials other than wood. Plastic material is most commonly used and all hive parts, including one-piece molded combs made from plastic, are available to the beekeeper. Generally, plastic equipment is as yet unproved, more expensive than wooden equipment, and being used to a very limited extent by the industry.

BREEDING AND GENETICS OF HONEY BEES

By John R. Harbo and Thomas E. Rinderer[1]

Domestic animals, such as chickens, cattle, sheep, pigs, and horses, have been selectively bred by human beings for thousands of years. Consequently, when modern breeding practices came into use, much selection had already been done; the modern animal breeder began with human-selected "breeds." The races of honey bees (such as Caucasians, Carniolans, and Italians) often are regarded as one would regard breeds of cattle or dogs. They should not be, for the honey bee races were not strongly controlled and bred by people and are much more variable than a breed of domestic animal.

The honey bee was not strongly selected by humans because basic bee reproduction was not understood until 1845. Without this understanding, very little could be done. In 1851, when this basic understanding was becoming widely accepted, Langstroth developed the movable frame hive. Suddenly beekeepers not only understood bee reproduction, they could also manipulate the hive and control the queen.

Controlling mating was the only obstacle remaining. Island isolation was one means, but it was of very limited value. Between 1860 and 1940, dozens of attempts were reported to induce queens and drones to mate in the confines of a jar, cage, tent, or greenhouse. Some claimed success, but the successes could not be verified or repeated. With the development of instrumental insemination as a practical technique in the 1940's, controlled bee breeding began.

Therefore, as people begin breeding bees, they enjoy the benefits of having a large and variable population with which to work. Breeders quickly discover that honey bees respond well to selection. In part, this is because human beings are just beginning to modify the bee through selection and controlled breeding.

Genetics of the Honey Bee

The family relationships within a colony of bees are different from other agriculturally important animals as a consequence of mating habits, social structure, and drones developing from unfertilized eggs. The honey bee colony found in nature is a complex family group, best described as a *superfamily*. This superfamily (fig. 1) consists of: (1) one mother queen, (2) several father drones present as sperm in a sperm storage organ (spermatheca) of the queen, and (3) the worker and drone offspring of the mother and fathers.

Within a superfamily are usually 7 to 10 *subfamilies*, that is a group of workers fathered by the same drone. Since all the sperm produced by a drone are genetically identical, each subfamily is composed of sisters that are more closely related than full sisters of other animals. Thus, workers belonging to the same subfamily, often called *supersisters*, have three-quarters of their genes in common by descent. They receive identical gametes from their father and, on the average, half-identical gametes from their mother.

Workers belonging to different subfamilies have the same mother but different fathers. They are half sisters and have one-quarter of their genes in common by descent. On occasion, brother drones mate with the same queen. In such instances, their subfamilies are related to each other as full sisters rather than half sisters. Through natural mating, such full sisters probably are uncommon.

Despite the complicated family structure, the basic principles of genetics still apply to bees. The chromosomes contain hereditary units called *genes*. The specific place on a chromosome where particular genes are found is called a *locus*. On rare occasions, a gene entering an egg or sperm has changed somewhat and will have a different effect than the original gene. The process of change is called *mutation*, and all the forms of a gene that might occur at a locus are called *alleles*.

[1] Research entomologist and research geneticist, respectively, Science and Education Administration, Bee Breeding and Bee Stock Research, Baton Rouge, La. 70808

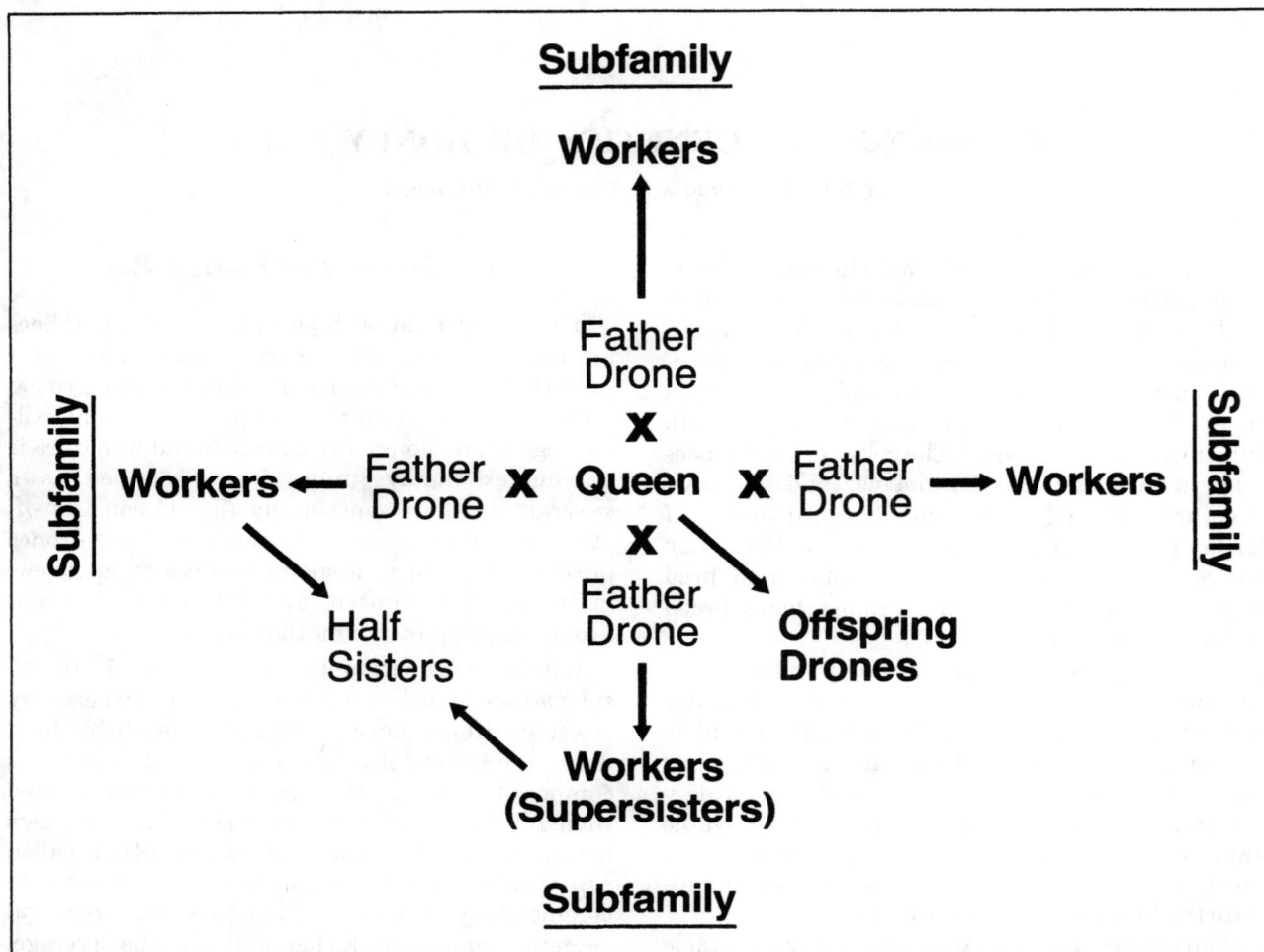

FIGURE 1.—A representation of a colony of bees as a genetic superfamily. The colony in the figure has four subfamilies, but there can be more or fewer. If two father drones are themselves brothers, the two subfamilies sired by them would be related as full sisters, rather than as half sisters.

Honey bee eggs hatch regardless of whether they are fertilized. The female bees—queens and workers—develop from fertilized eggs that contain 32 chromosomes. These 32 chromosomes consist of two sets of 16, one set from each parent. Hence, female bees are said to be *diploid* in origin. The males (drones) develop from unfertilized eggs which contain only one set of 16 chromosomes from their mother. Drones are thus *haploid* in origin. This reproduction by the development of unfertilized eggs is called parthenogenesis

Since queens and workers have paired chromosomes, they carry two alleles for each gene, one on each member of the pair. If both alleles are of the same type, the condition is homozygous; if they are different, the condition is *heterozygous*.

In some heterozygous circumstances, one allele will mask the expression of the other and is said to be *dominant*. The allele which has its expression masked is said to be *recessive*. Drones can carry only one type of allele because they are haploid; thus, they are called *hemizygous*.

At one time parthenogenesis was considered to be the basis of sex determination in bees. The theory was that a chromosome dosage effect occurred such that two sets of chromosomes resulted in females and one set resulted in males. While this is a reasonable explanation, this theory is now known to be untrue.

Research workers investigating apparent low egg viability in inbred lines discovered that sex in bees is determined by the alleles at a single locus.

If an egg is a heterozygote at this locus, it will develop into a female. If it is homozygous or hemizygous, it will develop into a male.

The apparent nonviable eggs found in the inbred lines were diploid eggs homozygous at the sex locus. Worker bees selectively remove and destroy homozygous diploid larvae from the comb just after they hatch. Research efforts have been made to rear these diploid drones to maturity with the hope of producing diploid sperm and triploid queens and workers. However, artificial rearing is a difficult procedure, and the resulting diploid drones had reduced testes and produced very little sperm.

While sex determination is genetically complicated, other characteristics can be even more complicated. Different combinations of alleles at a locus result in different expressions of characteristics. Alleles at other loci also can affect a characteristic. All these different events result in complex genetic systems which produce a wide variety of character expression in bees. For example, worker bee response to isopental acetate (a component of alarm pheromone) was estimated to be influenced by at least seven to eight genes. This variety is some of the raw material necessary for the genetic improvement of bee stocks.

As with other animals, variety in bees is further increased by events that occur when a female (queen) produces an egg. During this time pairs of chromosomes in cells that are destined to become eggs exchange segments. Further in the process, the chromosome number of germinal eggs is halved. This process results in a haploid egg, with chromosomes having a new combination of alleles at the various loci.

Unfertilized, the egg will develop into a drone which will produce sperm. The processes of recombination of alleles and reduction of chromosome number do not occur in drones. All the sperm cells produced by a drone are genetically identical. They are identical to each other, and they are identical to the chromosomes in the unfertilized egg that developed into the drone.

Communicating Pedigrees

Like other animal and plant breeders, bee breeders need a simple format for communicating ancestry and breeding plans. A pedigree format usually is a standardized diagram, simply showing a line from the father and one from the mother to one or more offspring. Because of the haploid drones, bee pedigrees are different.

To properly discuss a pedigree, two terms must be defined. These are *gamete* and *segregation*. An animal gamete is an unfertilized egg or a sperm cell containing half the chromosomes needed to produce a worker or queen. Segregation is the random sorting of paired chromosomes to produce gametes. In most animals, segregation occurs in the ovaries and in the testes. In bees, segregation occurs only in the ovaries of queens.

Therefore, in honey bees, all new gametes originate with a queen. We say "new" gametes because drones propagate only existing gametes. The drones then have two reproductive functions: first, they convert and extend the queen's female gamete (the single unfertilized egg that develops into a drone) into about 10 million identical male gametes (sperms). Second, they serve as a vehicle to move the propagated gametes to the queen (the act of mating).

In bee reproduction, then, the female progeny receive one gamete from the queen that produced the egg and the other gamete from another queen (via drone conversion of the gamete to a sperm cell). Thus, a bee pedigree contains only females. So instead of using the traditional circle to represent females and the square to represent males, the bee pedigree in figure 2 has only circles (ovals), for only females (or queens) need to be recorded.

Stock Propagation and Maintenance

Controlled Mating

Some degree of controlled breeding has been practiced by queen producers for over 75 years. During that time, beekeepers had the capability of producing hundreds of queens from a selected colony rather than relying on natural supersedure or swarming. Thus, the female line was controlled.

Controlling the mating has been possible only by establishing isolated mating yards or through instrumental insemination. Isolated mating yards have two major shortcomings: (1) Absolute control of matings is difficult to achieve because a queen can mate with drones that are up to 5 miles away, and (2) one isolated mating yard is needed for every drone line used in a breeding program. Mating yards usually are not used for breeding stock, but rather for production queens where absolute control of matings is not quite as critical.

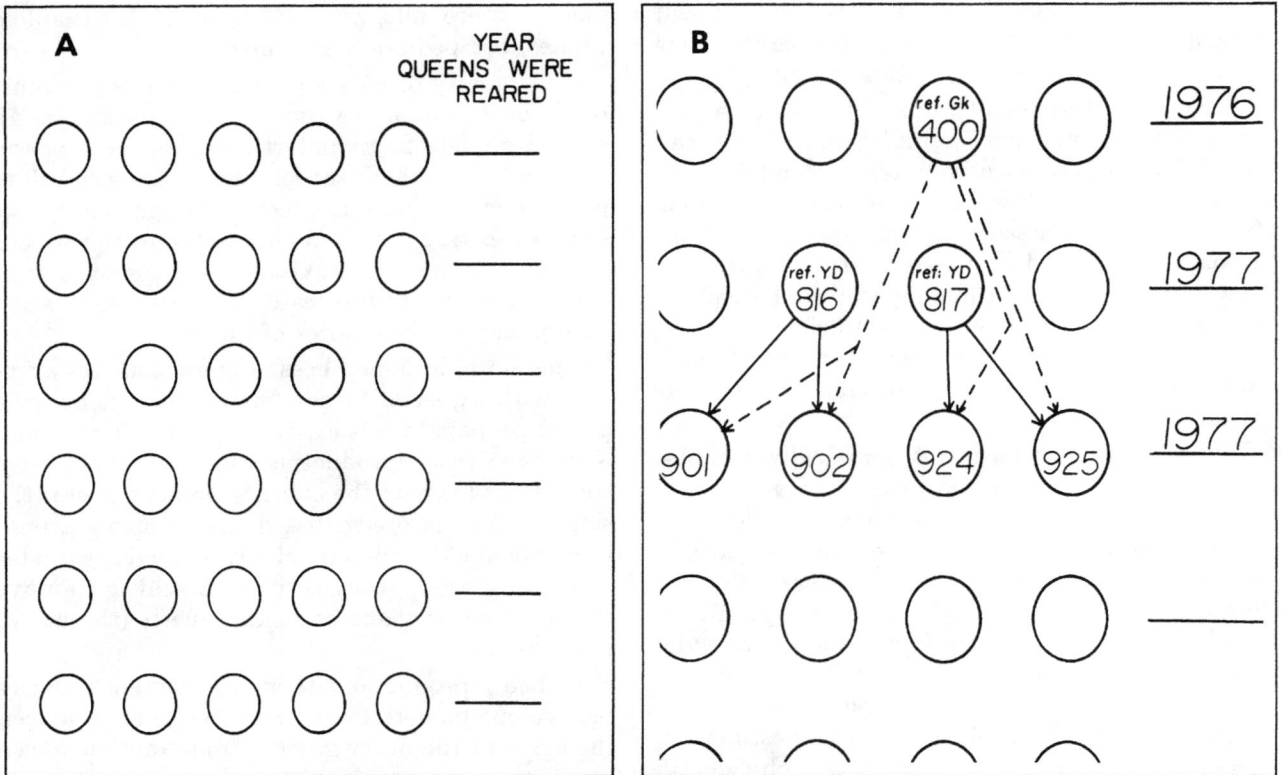

FIGURE 2.—A system for communicating bee pedigrees. *A* is a suggested format for a bee pedigree as printed on a sheet of paper; *B* uses the format of *A* to communicate some simple crosses. Each oval represents an individual queen and each line represents a gamete; dashed lines indicate male gametes (sperm) and solid lines indicate female gametes (eggs). The date on the right is the year that the queen or queens in that horizontal row were produced. Numbers in the ovals are for the breeder's convenience; they may be a queen number, a colony number, a breeding code, or a reference to another pedigree sheet. Any number of lines or gametes may generate from an oval, but each individual (oval) has either zero or two lines (one dashed and one solid) going to it. If a queen (oval) has no lines going to it, she is either unpedigreed or pedigreed on another sheet. Interpreting *B*: Drones from queen number 400 were mated to queens 816 and 817. Queen 816 was inseminated with semen from a single drone. Daughters subsequently were produced: 901 and 902 from 816, 924 and 925 from 817. The forked dashed line going to 901 and 902 indicates that the male gamete (sperm) going to queens 901 and 902 came from the same drone. Therefore, at least half the genes in queen 901 are identical with those of queen 902. The most common way a breeder would know this is a record of having inseminated queen 816 with a single drone. Queens 924 and 925 may or may not have a gamete in common.

Controlled breeding through instrumental insemination has been well established since 1947 and has solved the controlled-breeding problem for bee researchers, but commercial queen producers rely primarily on natural mating.

Recently, a few commercial queen producers have tried instrumental insemination for mass production of queens. Frequently, their reason for using it has been to eliminate mating nucs (small queen-mating colonies), rather than to make specific matings.

Instrumental insemination is, in itself, not a complicated process. Simply stated, it is a mechanical transfer of semen from the drone to the oviduct of the queen. This transfer can be made with any one of many designs of insemination stands and syringes available. All of them use carbon dioxide (CO_2) as an anesthetic to keep the queen still, use a device to hold the queen in position, and use some type of syringe to collect the semen and discharge it into the queen. Probably the most popular apparatus is that developed by Mackensen (fig. 3). Usually, one can become proficient at instrumental insemination, after practicing with 50 to 100 queens. Thereafter, the major problems and the major workload involve drone rearing, holding adult drones to maturity (about 2 weeks), queen storage, coordination of queen and drone

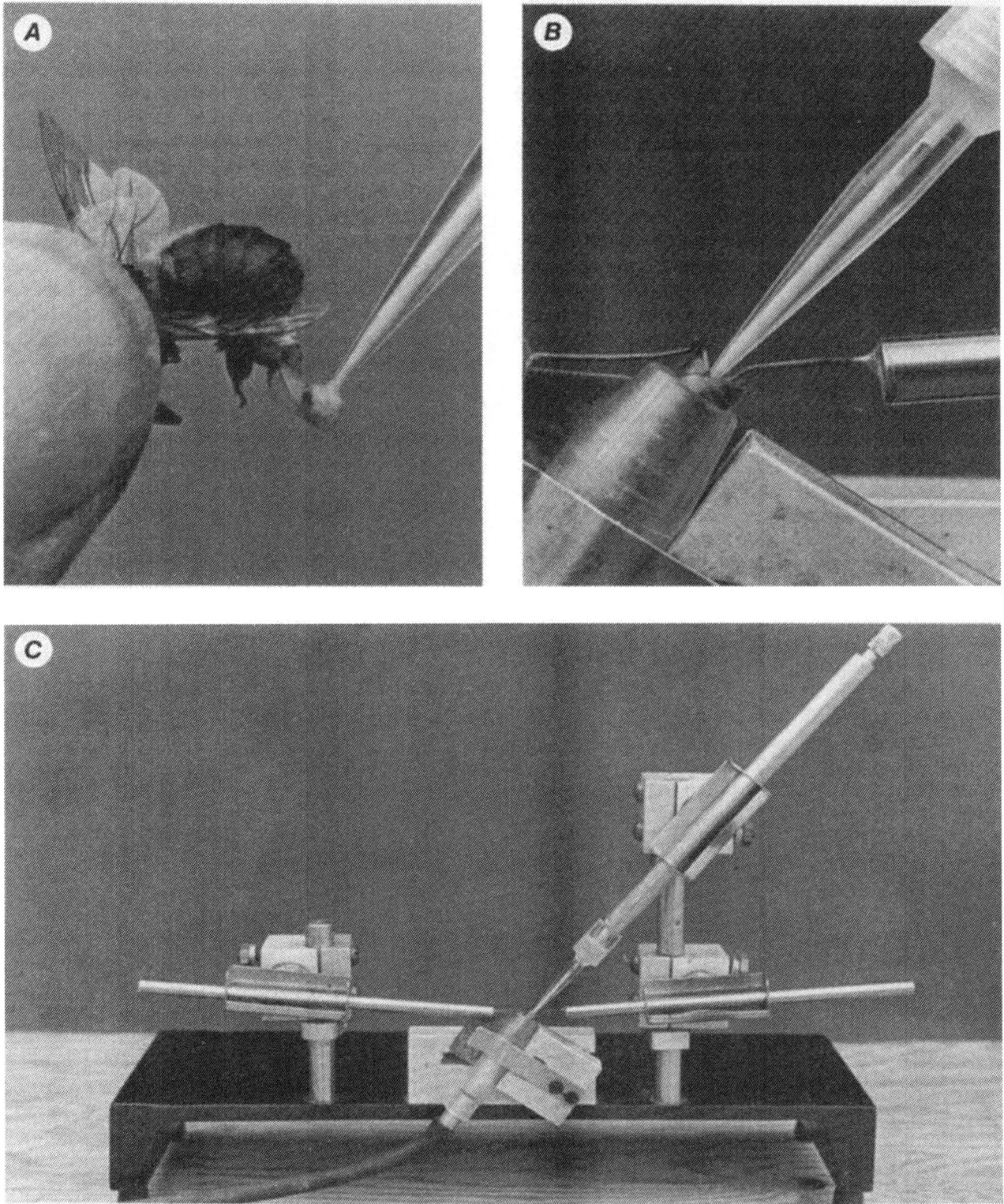

FIGURE 3.—Semen collection and queen insemination with the Mackensen insemination apparatus: *A* and *B* show collecting semen from a drone and the insemination of a queen; *C* is a wider view of the apparatus.

production, queen introduction, and record-keeping.

Instrumentally inseminated queens differ from naturally mated queens in their initial egg laying; instrumental insemination does not stimulate egg laying as does natural mating. If it were not for the carbon dioxide anesthetic given to queens during the following insemination, an instrumental-inseminated queen would begin laying eggs no sooner than a virgin queen that was not allowed to leave the colony to mate. The usual procedure is to render queens unconscious with CO_2 for 10 minutes on each of the 2 days following insemination.

Germplasm Storage

Germplasm is the hereditary material that can produce new individuals. In honey bees, this includes eggs, sperm, and tissue that can potentially produce eggs or sperm. Since every breeding program needs to keep certain stock for current and future use, the problem becomes one of either storing it (as with seeds on a shelf) or continually propagating it.

At the present time, honey bee germplasm is kept primarily through propagation. Thus, germplasm is usually in the form of mated queens—their ovaries and the sperm in their spermathecae. Alleles are lost gradually through inbreeding, so each generation of propagation reduces the variability of the germplasm slightly.

To avoid this loss and the labor involved in propagation, attempts have been made to store honey bee germplasm. Among the possible candidates for storage (eggs, larvae, pupae, virgin queens, sperm), sperm storage has been the most successful. Sperm stored less than 2 weeks at nonfreezing temperatures seems to be as viable as fresh sperm, but longer storage results in fewer sperm reaching the spermathecae. Although inferior inseminations result from sperm stored in liquid nitrogen ($-196°C$), nitrogen shows great promise for long-term storage where survival of the germplasm is the major concern.

Mutations

More than 30 specific visible mutations have been described in bees, and a number of these are maintained by research laboratories. Generally, these mutations produce a striking effect, and the majority have been easily observed by their discoverers. Many other mutations might occur in bees that also cause subtle changes yet to be observed. Known mutations affect the color, shape, and presence of eyes, the color and hairiness of bodies, the shape and size of wings, and nest-cleaning behavior.

Probably because of their distinctive appearance, most of the honey bee mutants thus far collected had variations in color of eyes. Various shades of white, tan, chartreuse, and red have been described and about 20 still are maintained. In addition to their value as curiosities, these mutants have value as scientific tools. For example, by studying various colors of eye mutants, the biochemical pathway for the production of eye pigments in honey bees was determined.

In addition to contributing to work on eye pigment biochemistry, mutants have been used as tools to investigate a variety of other questions. Resistance to American foulbrood, mating behavior, sex determination, pollination activity, fertilization technology, sperm storage, population dynamics, longevity, and bioacoustics all have been explored with experimental designs utilizing bees identifiably different because of mutations they carry. Because of this history of usefulness and further potential applications, it is desirable for the scientific community to maintain a number of mutations. Newly discovered mutations may have special applications in science; therefore, beekeepers could help by reporting mutations they observe to a research laboratory.

Most mutations are recessive. Mutations, therefore, are often first observed in drones, for drones are haploid and do not mask recessive genes. A mutation might occur in a single drone in a colony or in many drones.

Gene Pool

Across the world, bees are quite diverse. Time, mutations, and selection pressures have resulted in populations of bees called races, somewhat isolated from each other, that excel for various combinations of characteristics. These combinations of characteristics are finely tuned for survival in specific local environments. Worldwide, the races of bees form the gene pool or genetic base available to bee breeders for stock improvement.

Since North America and South America lacked native honey bees, European settlers imported them. Early importations were the brown bees common to northwestern Europe. Through time, beekeeping developed as an industry in North

America and beekeepers, happy with some characteristics of the European brown bee and unhappy with others, made further imports. Prominent among these imports were bees from other European areas. However, bees also were brought from Africa and Asia. The search by beekeepers for better bees led to a wide variety of genetic material being brought into America until 1922, when importing adult bees was banned to prevent the mite *Acarapis woodi* (Rennie), the cause of acarine disease, from entering the country. Eggs and semen, however, were still imported but at a much reduced rate. After 1976, importing any honey bee germplasm was banned by Public Law 94–319, except under authorization from the U.S. Department of Agriculture.

The bees presently in the United States are the result of free-mating crosses of the various imports. Most probably, racially "pure" stocks no longer exist in North America. Rather, this new genetic mix of bees can best be termed American. By the same token, exports from the United States and crossbreeding have influenced the nature of bees abroad.

The great virtue of our past imports is that we can breed highly desirable bees from the many varieties of stocks that we now have on the continent. Yet, some people still want to import more stock. Usually, the first thought that comes to mind when improved stock is desired is to import. It seems like a simple solution. Importing, however, should not be used as an easy substitute for a selection program. If used, importing would be only preliminary to a selection program, as an effort to expand the genetic base from which to select.

Thus, further stock imports are of questionable benefit. Such imports may result in acarine disease, may be themselves undesirable, or may combine with local stocks to produce undesirable hybrids. Many past imports of "select stock" proved to be poor or even undesirable in North America. In addition, imported bees may have originated from bees exported by American queen producers. For those interested in improving bee stock, therefore, it is probably best to select from the plentiful gene pool already available in North America.

Stock Improvement

Using improved stocks of bees is an effective way to improve the productivity of a beekeeping operation. Regardless of the stock of bees used,

basic operational expenses will remain much the same.

Success in improving bee stocks is a reachable goal. As we have seen, there is great variation in bee stocks available to North American bee breeders. This variation is the raw material used by bee breeders. Working with the tool of selection, bee stocks can be molded to show high performance for desired characteristics.

Selection Methods

Describing the desired stock

The first task of a bee breeder is to describe in rather specific terms what characteristics are desired in the bee stock to be produced. Almost certainly a number of characteristics will be listed. Generally, desired characteristics will relate to the production needs of a group of beekeepers who are in similar localities or have similar needs. Desirable characteristics might include fast spring buildup, intensive honey production, frugal and strong overwintering ability, disease resistance, and good handling qualities. A different list might emphasize heat tolerance and pollination activity.

A knowledgeable bee breeder will be careful to be only as specific in his stock descriptions as good information permits. Unless scientific proof is developed to the contrary, physical characteristics such as color, size of bees, and shape of wings are poor choices. Generally, if such characteristics are important, they will be selected and improved automatically along with more general characteristics such as honey production or disease resistance.

A knowledgeable bee breeder also will set reasonable goals. Some characteristics, such as frugal use of winter stores and early strong buildup, are not likely to be highly compatible. A list of a few well-chosen characteristics is more likely to be achieved than a longer list.

Overall, there is a need in the beekeeping industry for a number of bee stocks, each having a collection of characteristics economically important to different segments of the diverse beekeeping community. No one bee stock can possibly be universally acceptable, and attempts to produce such a stock would prove fruitless. Thus, communication between an individual bee breeder and the beekeepers using the breeder's stock is important. This communication will help the breeder decide which characteristics to emphasize in the breeding program.

Measuring superior breeding stock

Once the breeding goal has been established by describing the desired stock, choices need to be made as to how the various characteristics will be measured. Although more precise ways to evaluate colonies may be devised in the future, at present the bee breeder must choose his stock from the on-site performance of colonies established in apiaries.

Test apiaries should be established with colonies arranged in an irregular pattern, with the colonies spaced as far apart as possible. Apiary sites that have trees, shrubs, or other such landmarks are valuable. These various precautions will tend to prevent the drifting of field bees. Other management procedures also should be reasonably uniform so that all colonies have an equal opportunity to perform. Management procedures should conform reasonably well to the management procedures used with production colonies.

Test colonies will be evaluated for the various characteristics to determine which colonies will be used as breeding stock. In all cases, beekeeping judgment will be brought to bear on the evaluation. The power to select more accurately the best breeders, however, will be enhanced if each colony is given a numerical score for each characteristic being evaluated. This will require the keeping of extensive records on colonies. Obviously deficient colonies can be left out of the recordkeeping to ease the load. Such records are particularly important when different evaluations, such as honey production and overwintering ability, are made at different times.

Once all the colonies have been evaluated, which will take 1 to 2 years, depending on the characteristics to be improved, breeder colonies can be chosen. The scores given to a single colony for all the various characteristics can be added to provide a single numerical score for the entire colony. Such scores can then be compared to select the best colonies available. More emphasis can be put on one or another characteristic by adjusting the scores given for that characteristic. For example, honey production may be scored on a scale of 0–20, while temper may be scored on a scale of 0–10. This arrangement would be used if honey production was considered twice as important as temper.

Breeding Methods

Line-breeding

The common method of breeding practiced by queen breeders is known as line-breeding. It can be defined as breeding and selecting within a relatively small closed population. The bee breeders' colonies constitute such a population to the extent that mismatings with drones outside their stock do not take place.

The general procedure in line-breeding is to rear queens from the best colonies. These queens are both sold as production queens and used to requeen the bee breeder's test colonies. The queens are allowed to mate with the drones present in the bee breeder's outfit at the time the queens are reared.

A number of variations can be made on this general procedure which would be of benefit. General control of drone brood in the majority of colonies, coupled with purposeful propagation of drones in a good number of more exceptional colonies, would improve the selection progress by controlling, to a limited degree, the male parentage of the stock. Of course, this procedure is used with the best success in areas where mating yards can be reasonably isolated.

In line-breeding, some inbreeding is inevitable. Its main effects are (1) fixation of characteristics so rapidly that effectiveness of selection for good qualities is reduced, (2) the stock loses vigor as a general consequence of inbreeding, and (3) the poor brood pattern from homozygous sex alleles. These effects can be lessened by using as many breeding individuals as possible for every generation.

To keep inbreeding at a minimum, one should rear queens from as large a number of outstanding queens as possible and requeen all the field colonies with equal numbers of queens from all the breeders. Each group of queen progeny is then considered a queen line and each year, after testing, at least one queen in each line is used as a grafting mother.

Despite these several precautions against inbreeding, stock may begin to show a spotty brood pattern and other symptoms of inbreeding. When this occurs, new stock must be brought into the operation. At least 10 virgin queens from each of several promising stocks should be mated with drones of the declining stock and established in apiaries outside the mating range of the beekeeper's queen-mating yards. They should be evaluated there to determine which stock(s) combine best with the deteriorating stock. Once this evaluation is made, the preferred stocks can be established as new queen lines.

In the 1930's, a 4-year selection project using simple line-breeding resulted in an increase in honey production from 148 to 398 pounds per colony. Two important features of this project were culling the poorer queens and grafting from the best queens.

Hybrid breeding

When inbred lines or races of bees are crossed, the hybrid progeny often are superior to either parent for one or many traits. This phenomenon is called *hybrid vigor* or *heterosis*. Hybrid bees have more heterozygosity in their genome than do inbred or line-bred bees. This heterozygosity is thought to be the basis for hybrid vigor.

Hybrid-breeding programs in bees are considerably more complicated than line-breeding programs. At the very least, three inbred lines must be combined so that both queens and their worker daughters are hybrids. An inbred queen mated to inbred drones will produce hybrid workers. However, the egg-laying qualities of the inbred queen probably would be inadequate. Therefore, there is a need to mate hybrid queens to inbred drones so that both queens and workers in production colonies are hybrids.

Four-line hybrids also are possible and commercially available. Such a hybrid may involve lines 1, 2, 3, and 4 and could be combined in the following way: An inbred queen of line 3 artificially mated to drones of line 4 is used as a grafting queen to produce hybrid (3 × 4) queens. These are allowed to mate naturally and are used to produce drones. Queens of line 1 are then mated to drones of line 2 and hybrid virgin queens (1 × 2) are reared from the mating. Production queens are produced from a cross of virgin queens (1 × 2) mated to the drone progeny from the 3 × 4 queens. Colonies produced by this cross will be headed by two-way hybrid queens, which will be uniform in appearance, whereas the worker bees will be four-way hybrids and variable in appearance, unless the color markings of the parent lines are very similar.

Comparative tests of hybrids have shown their superiority. Increased productivity of 34 to 50 percent over the average of line-bred strains has been reported. Segregation and random mating in the generations following hybridization are likely to result in queens that are no better than the average supersedure queen. Hybrids are an end product, and to make best use of them it is necessary to requeen every year.

Whatever the specific choice of breeding scheme, hybrid breeding requires the use of instrumental insemination and careful recordkeeping. As a consequence, few bee breeders have undertaken the entire operation of a hybrid program. However, many have become involved as producers of hybrid queens with the breeding stock supplied by an outside source.

ACKNOWLEDGMENT

We thank Dadant & Sons, Hamilton, Ill., and G. H. Cale and W. C. Rothenbuhler for permission to modify and use figure 1, which appeared in *The Hive and the Honey Bee.* James Baxter, biological technician in our laboratory, prepared figure 1.

References

CALE, G. H., JR., and W. C. ROTHENBUHLER.
  1975. GENETICS AND BREEDING OF THE HONEY BEE. *In* The Hive and the Honey Bee, p. 157–184. Dadant & Sons (ed.) Journal Printing Co., Carthage, Ill.

DZIERZON, J.
  1845. GUTACHTEN UBER DIE VON HR. DIREKTOR STÓHR IM ERSTEN UND ZWEITEN KAPITEL DES GENERAL–GUTACHTENS AUFGESTELLTEN FRAGEN. Bienen-Zeitung (Eichstatt) 1:109–113, 119–121.

LAIDLAW, H. H., JR.
  1977. INSTRUMENTAL INSEMINATION OF HONEY BEE QUEENS. *In* Pictorial Instructional Manual. 144 p. Dadant & Sons, Inc. Journal Printing Co., Carthage, Ill.

MACKENSEN, O.
  1951. VIABILITY AND SEX DETERMINATION IN THE HONEY BEE ("APIS MELLIFICA" L.). Genetics 36(5):500–509.

——— and K. W. TUCKER.
  1970. INSTRUMENTAL INSEMINATION OF QUEEN BEES. U.S. Department of Agriculture, Agriculture Handbook No. 390, 28 p.

POLHEMUS, M. S., J. L. LUSH, and W. C. ROTHENBUHLER.
  1950. MATING SYSTEMS IN HONEY BEES. Journal of Heredity 61(6): 151–155.

RINDERER, T. E.
  1977. A NEW APPROACH TO HONEY BEE BREEDING AT THE BATON ROUGE USDA LABORATORY. American Bee Journal 117(3):146–147.

ROTHENBUHLER, W. C., J. M. KULINCEVIC, and W. E. KERR.
  1968. BEE GENETICS. Annual Review of Genetics 2:413–438.

QUEENS, PACKAGE BEES, AND NUCLEI: PRODUCTION AND DEMAND

By Kenneth W. Tucker [1]

The production of queen bees, package bees, and nuclei provides for the establishment of new colonies, the replacement of dead colonies, or the rejuvenation of ongoing colonies. It renders these objectives as planned management, instead of the haphazard and often ill-timed replacement of colonies by swarming and of queens by swarming or supersedure.

Most queens, packages, and nuclei are produced by beekeepers who concentrate on this specialty of beekeeping. Most beekeepers who manage bees for the production of honey and wax or for pollination leave the propagation of queens and bees to a specialist for several reasons. Probably the most important reason is that the specialist in queens and bees can produce them at less expense than could other beekeepers. The specialist also requires considerable experience in many detailed rearing techniques, in comparing breeder queens to ideal standards of performance, and in testing for uniform performance of the daughter queens. The specialty also requires considerable investment in equipment used just for queen and bee production.

The annual dollar value of queen and bee production for 1975 was probably about $10 million to $15 million. California, where county-by-county production statistics are available, accounted for about half this production. For Texas and the Southeastern States, the estimate was based on the guess that about half the queens and a third of the package bees were produced there. This estimate indicated a production of about 1 million queens and 500 tons of package bees but did not include a complete accounting for queens, bees, and nucs (nuclei, small colonies) used within the same beekeeping operation. Partial figures for large migratory bee operations indicated a production of perhaps 100,000 valued at about $2 million.

As an industry within an industry, the queen and bee business operational expenses and earnings become the operational expenses of other beekeepers. Some large honey producers and pollination contractors have integrated queen and bee rearing into a "vertically" enlarged beekeeping operation. Typically, such operations locate in mild winter areas for queen and bee production during winter and spring, then locate in cold winter areas for honey production or pollination in summer.

Queens

Queens usually are sold as young mated adult queens that have been laying eggs for only a few days. These queens most often are reared from 12- to 24-hour-old worker larvae, transferred ("grafted") from worker comb into specially prepared queen cell cups. The developing queens are reared in either queenless colonies or next to young brood in part of a queenright colony from which the queen is excluded. Completed queen cells, when within a day of the queen's emergence as an adult, are placed individually into small, queenless colonies of bees, called mating nucs. About 2 weeks later, the young queens will have mated and be laying. For shipment, each queen is placed into a mailing cage supplied with candy along with 7 to 10 young worker bees from the mating nuc (fig. 1).

Package Bees

Package bees usually are sold caged in 2- or 3-pound units, along with a caged laying queen and a pint can of sugar syrup. Bees for packaging usually are shaken from the brood combs of the upper part of strong colonies, so that mostly young adult bees will be included. The bees are shaken through a funnel into the packages or into a "shaker box" until about 10 pounds of bees have accumulated; then several packages are filled from the shaker box. Cages for shipping packages are made of wood and screen, in a way that com-

---

[1] Research entomologist, Science and Education Administration, Bee Breeding and Bee Stock Research, Baton Rouge, La. 70808.

bines lightweight, sturdiness, and a maximum of screened area through which the bees can ventilate their cluster (fig. 2).

Nucs

Nucs or nuclei are small colonies with queens. They usually are composed of three to five combs of bees and brood with a laying queen. These may have been the mating nucs for the resident queens, or queens mated elsewhere could be introduced into nucs newly assembled from strong colonies. Nucs intended to develop into full-strength colonies are made up in equipment of the same size as the full-strength colony, in contrast to most queen-mating nucs, which have much smaller combs (fig. 3).

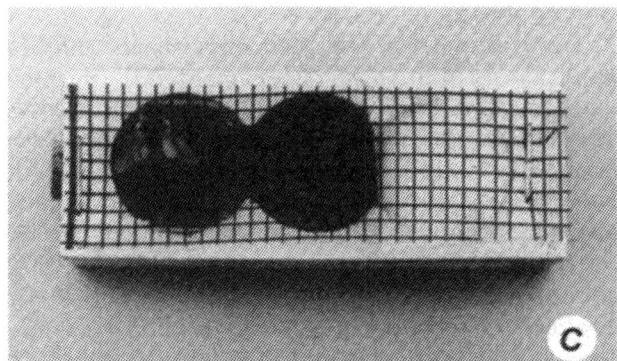

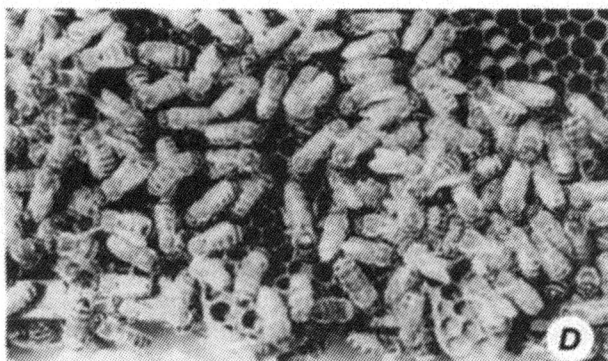

FIGURE 1.—Queen production: *A*, Completed queen cells; *B*, mating nucs in queen mating yard; *C*, young laying queen on a nuc comb; *D*, queen and attendant worker bees in shipping cage.

FIGURE 2.—Package bees: *Left*, Single package; *right*, packages crated for shipment.

FIGURE 3.—Popular comb sizes for nucs. Smaller size is mating ("baby"), size 6¼ × 8⅛ inches; larger, standard Langstroth comb, 9⅛ × 17⅝ inches.

Management for Queen and Bee Production

Queens are best reared and bees and brood most efficiently produced under those circumstances which accompany the spring swarming season: colonies near peak population, an abundance of nectar and pollen, and a readiness to rear drones in abundance. The season of abundance can be started earlier and extended later by feeding sugar syrup and pollen substitutes when nectar and pollen flows are inadequate. For maximum production of queens, drones, and bees, conditions of abundance should be maintained for as much as 2 to 2½ months before the anticipated date of queen production and continued as long as queens are still to be mated. Drone production is the most sensitive to nutritional inadequacies and should be used to gage rearing conditions. By early and continued feeding, the rearing season may be advanced by a month and production of bees for stocking mating nucs and for packages increased.

Demand for Queens and Bees

The demand for bees and queens is based on the use made of them. These uses usually are the start of new colonies and the continuance of established colonies.

New colonies may be started in a variety of ways. A 2-pound package with a laying queen installed onto drawn comb and provided with adequate honey or syrup and pollen or pollen

supplement should be expected to develop into a strong colony in about 12 weeks (four brood cycles). A three- or five-frame (9⅛-inch depth Langstroth) nuc with a queen should be expected to develop to the same strength in 9 weeks (three brood cycles) because it has brood already present. Either the package or the nuc option may be made starting with only a shipped queen, but with the bees and brood supplied from other colonies managed by the recipient of the queen. Of these two options, the nuc (also called a divide or split) probably has been used most, with the bees and brood taken from the strongest overwintered colonies. A few northern beekeepers shake their own packages.

For the continuance of established colonies, usually only a queen is needed. These queens are used to replace older queens which are no longer laying well or to change the nature of the bees of a colony that may be unproductive, overwinter poorly, sting too much, use too much propolis, or have any other characteristic the beekeeper thinks undesirable.

The demand for bees and queens also reflects the ways in which the recipients expect the bees and queens to fit their management. The recipient beekeepers must make decisions based on economics and on the seasonal cycle of weather and honey flows in their localities. These decisions include whether (1) to operate with perennial (overwintering) or annual (bees killed in late summer or fall and the hive restocked in the spring) management (which depends upon whether the beekeeper expects surplus honey from new colonies in the same season or not until the next year), (2) surplus honey flow or pollination is early, mid-season, or late, and (3) to establish new colonies or replace queens.

These management decisions by the recipients of queens and bees have dictated two aspects of the queen and bee industry: the shipping season and the geographic location of the industry.

The relation between management options and shipping season is summarized in table 1. Currently, and for the past 50 years, the shipping season for most queens and bees has been early spring, from March to May. The reason for this has been the dual demand for bees to replace winter losses and a very strong demand from annual management without overwintering. But with less annual management and more overwintering, it

TABLE 1.—*Demand for queens and package bees or nucs related to bee management and to timing of honey flows or crop pollination*

Management		Objective	Requirement	Time of year to receive queens and bees [1]		
Type	Time of flow or pollination			Early spring	Late spring to early summer	Late summer to early fall
Perennial____	Late spring–early summer.	Replace queens_____	Laying queens___ ____	P	O	P
		Replace and/or increase colonies.	Packages or nucs with queens.	P	O	
Perennial____	Fall_____	Replace queen_____	Laying queen_____	P	P	P
		Replace and/or increase colonies.	Packages or nucs with queens.	P	P	
Annual [2]____	Late spring–early summer.	Replace colonies_____	Packages or nucs with queens.	P	_____	
Annual [2]____	Fall_____	Replace colonies_____	Packages or nucs with queens.	O	P	

[1] P-prepared time, O-useful optional time.

[2] Bees restocked each year and killed after honey flow or crop pollination.

seems possible that the shipping season may spread out from the spring concentration.

Because of the heavy demand for queens and bees in early spring, the queen and bee industry has become concentrated in the mild winter areas of the country. Thus, most queens and bees currently are produced in Georgia, Florida, Alabama, Mississippi, Louisiana, Texas, and the Sacramento Valley of California.

Shipment

The shipment of bees and queens between the producer and recipient relies on rapid transportation and careful handling en route. Queens are shipped in small wood and wire-screen cages, along with 7 to 10 attendant worker bees and some candy, via airmail as far as the other side of the world. Package bees are shipped largely in the recipients' trucks, where they receive optimal care. For package bees, air freight is feasible but expensive; railway express, once the most preferred by the industry, now has too few routes and too many handling problems; parcel post is feasible for small shipments, but handling has been variable. The shipment of nucs, as with larger colonies, is best attended to by experienced beekeepers, who know that bees need water and cannot stand much heat: therefore, they are best shipped on the recipient's truck.

Looking Back

Planned queen rearing began in the mid-19th century with the advent of the movable comb hive. Besides rearing queens from swarm or supersedure cells, queens were reared from eggs or very young larvae in bits of worker comb taken from the breeder colony and transferred to a strong queenless colony. (That queens could be reared in this way was already known by 1771, but was facilitated by the use of the movable comb hives.) Several methods of planned queen rearing were developed by the 1880's, of which the most convenient and widely used to this day is that outlined above. Controlled mating by instrumental insemination, developed between 1925 and 1945, has been used by a few queen producers for breeding and lately is beginning to be used for mass-produced queens.

The shipment of queens in small cages, with attendant worker bees and food, developed apace with queen-rearing techniques. Early shipments of queens caged with a piece of comb honey and and a bit of water-soaked sponge succeeded by railroad transit for up to a few thousand miles. But distant shipments, as from the Eastern United States to New Zealand, took up to 63 days by rail and ship, and for these, adding a small piece of emerging worker brood to the cage enhanced the chance of survival of at least some queens of a shipment. By the 1880's, a fondant-type candy was being used to replace the comb honey, in response to the express carriers who found the honey-supplied shipments messy. Since the advent of reliable airmail, shippers have sent queens to all parts of the world with excellent prospects of survival.

The reasons for the demand for queens have changed since the inception of planned queen rearing. The earliest demand reflected a desire by North American beekeepers to change their stock from the existing dark bees of Northern Europe to the yellow Italian stock. This change of stock extended over an 80-year period for the country as a whole and was essentially completed by 1940. During this time, the demand for changed stock gradually diminished but is still a motive for buying queens. Demand based on replacement of poor and failing queens has been steady over time. Beginning in the late 1930's and continuing, there has been a strong demand for queens with packages for annual management on the northern honey flows.

The use of package bees was first conceived and tried in the late 1870's but really developed during the second decade of the 20th century. The demand in earlier days was sporadic and substantial only during springs following widespread heavy winter losses of colonies in the north. Volume trade in package bees dates from 1913. From that time until the 1950's, package bees were shipped with reasonable success by railway express. With the gradual reduction in rail passenger service and with the construction of high-speed highways, however, recipients of package bees adjusted by carrying package bees north by truck.

The earliest demand for package bees was to start new colonies, including replacing those that died during winter, without the expectation of a full crop of honey the first year. But with the demonstration in Manitoba during the 1920's that reasonably large honey crops could be expected during the first year from spring-package colonies, a new type of annual management developed over the North-Central and Northwestern United States and in Canada. Package colonies established early in the northern spring build up to full strength before the summer honey flow and produce a crop of surplus honey, after which the bees are killed in late summer or fall. The demand for package bees used in annual management is still substantial, but in recent years perennial management in the north has been competitive with annual management, leading to a lower demand for package bees.

Nuclei with queens have been offered for sale since at least the 1870's. Survival of shipments in early days by rail, barge, or ship was uncertain even when the bees were accompanied by a skilled beekeeper. Only in the last 20 years has volume shipment of nucs over large distances been feasible, when they are carried on the recipients' trucks over high-speed highways.

The location of the queen and bee industry has changed over the years since the 1860's. Pioneers of the industry were located mainly in the northeastern United States. By the 1880's, queen producers were located across the country, even in very cold winter areas. Before the advent of reliable shipment of package bees, the greatest demand for queens was during late summer and fall, so that northern-reared queens were competitive. Dating from the 1870's, a few southern queen producers sold queens in early spring, but the shift of the larger part of the queen and bee industry to mild winter areas did not start before the 1910's. By the 1930's, the bee industry was located mostly in the south and in California, where it is today.

Looking Ahead

It is intriguing to guess what lies ahead for the queen and bee industry. Any changes in rearing technology, handling, and shipping of queens and bees, or in management objectives by customers, can be expected to influence this industry.

For rearing technology, future changes should be directed toward reducing the amount of detailed manual skills, the number of colonies operated per numbers of bees and queens produced, and the amount of travel within a queen and bee operation. An innovative approach to grafting may yield a system to graft 10 to 20 cells as fast as one cell is grafted now. Instrumental insemination may yet become superior to natural mating and, if so, outyards for natural mating and mating nucs could be supplanted by many fewer nursery colonies. Driving or attracting bees into cages by using repellants or attractants may replace shaking. The advent of a protein diet superior to pollen in nutrition and attractiveness to bees could lead to "feedlot beekeeping" and the elimination of outyards for queen and bee production.

Methods of shipping bees and queens probably can be improved. A diet for bees in transit better than the currently used sugars may give better survival. Controlling the climate for queens and bees in transit may be used more extensively in the future than now, building on promising systems now in use.

Finally, future changes in management objectives of customer beekeepers may change the nature of the bee and queen industry. Probably the most important change, perhaps already underway, is the return from annual to perennial management with overwintering. Another conceivable change is the exploitation of surplus honey from very early spring flows, formerly considered "buildup" flows. Both these changes may shift the season of demand for queens and bees to summer and fall. If this happens, volume queen and bee production may again shift northward.

References

LAIDLAW, H. H., and J. E. ECKERT.
  1962. QUEEN REARING. 165 p. University of California Press, Berkeley and Los Angeles.
PELLETT, F. C.
  1938. HISTORY OF AMERICAN BEEKEEPING. 393 p. Collegiate Press, Ames, Iowa.
ROBERTS, W. C., and W. STANGER.
  1969. SURVEY OF THE PACKAGE BEE AND QUEEN INDUSTRY. American Bee Journal 109:8–11.
WEAVER, R. S.
  1969. PAMPERED PACKAGE BEES. American Bee Journal 109:49.
YORK, H. F., JR.
  1975. PRODUCTION OF QUEENS AND PACKAGE BEES. Chapter XIX *in* The Hive and the Honey Bee. 740 p. Dadant & Sons, Hamilton, Ill.

MANAGING COLONIES FOR HIGH-HONEY YIELDS

By F. E. MOELLER[1]

Colonies of bees existing in the wild, away from the control of human beings, will produce small surplus crops of honey above their requirements for survival. Such surplus will vary, depending on the region or locality, but will seldom exceed 25 to 30 pounds. In the same area and with the same nectar resources, colonies properly managed will produce surplus honey crops exceeding 100 pounds. Intensive two-queen colony management often can result in surplus crops of 300 pounds or more with the same resources available. The key to these differences is management.

Proper management employs practices that harmonize with the normal behavior of bees and brings the colony to its maximum population strength at the start of the bloom of major nectar-producing plants. Management practices are similar in basic principle wherever bees are kept and vary only as regards timing for the desired nectar source of the region or locality concerned.

Honey bee biology is constant. Bees respond to their environment as temperatures and food supplies are changed. Beekeepers, in managing or manipulating colonies, are merely facilitating normal biological colony changes to suit their purpose. They can accelerate brood rearing by pollen feeding and hive manipulation, or they can crowd or restrict colony activity by certain other manipulations. Responses of the colony, wherever it is kept, are predictable. Thus, the basic handling, management, and manipulation of bees are universally similar, varying only as to localities and the timing of bloom of the major nectar and pollen plants.

Regardless of the type of hives or equipment used, proper management aims at providing colonies with unrestricted room for brood rearing, ripening of nectar, and storage of honey, plus provision of adequate food requirements, both pollen and honey, for the time of year concerned. Swarm-

ing is minimized and the storing instinct encouraged when proper management is used.

Preparing Colony for New Season

In the temperate regions of the Northern Hemisphere, August to October is the time when beekeepers prepare their colonies for the coming year. This is when the major honey flows are usually past and the bees must be made ready for the coming winter.

All queens of questionable performance with only a small amount of brood of irregular pattern (fig. 1, A) should be replaced. Frequently, the bees of the colony will replace or supersede queens of subnormal performance even before the beekeeper senses a problem. Some queens may be satisfactory in their second year; queens less than a year old are usually best.

To requeen a colony, certain principles of queen acceptance must be borne in mind: (1) Strong colonies more reluctantly accept a queen than weaker ones, (2) temperamental bees are more reluctant to accept a new queen than gentle bees, (3) young bees accept a queen more readily than older bees, (4) the colony to be requeened should first be made queenless, and (5) the queen to be introduced should be in egg-laying condition.

There is less risk in requeening a colony by giving it a laying queen with some of her own brood and bees than by giving it a queen in a shipping cage. A new or valuable queen should first be introduced into a small colony or divisions of one in a queen-shipping cage. After she is laying, the small colony can be united with a large one.

A drone-laying queen can be replaced if she is discovered while the colony is still strong. If the colony is weak, the bees should be removed and the equipment added to another colony.

Assuming colony conditions and the condition of the queen are favorable, the effect of environmental or working conditions and the time of year are factors that affect queen acceptance. Best

[1] Research entomologist, Science and Education Administration (deceased).

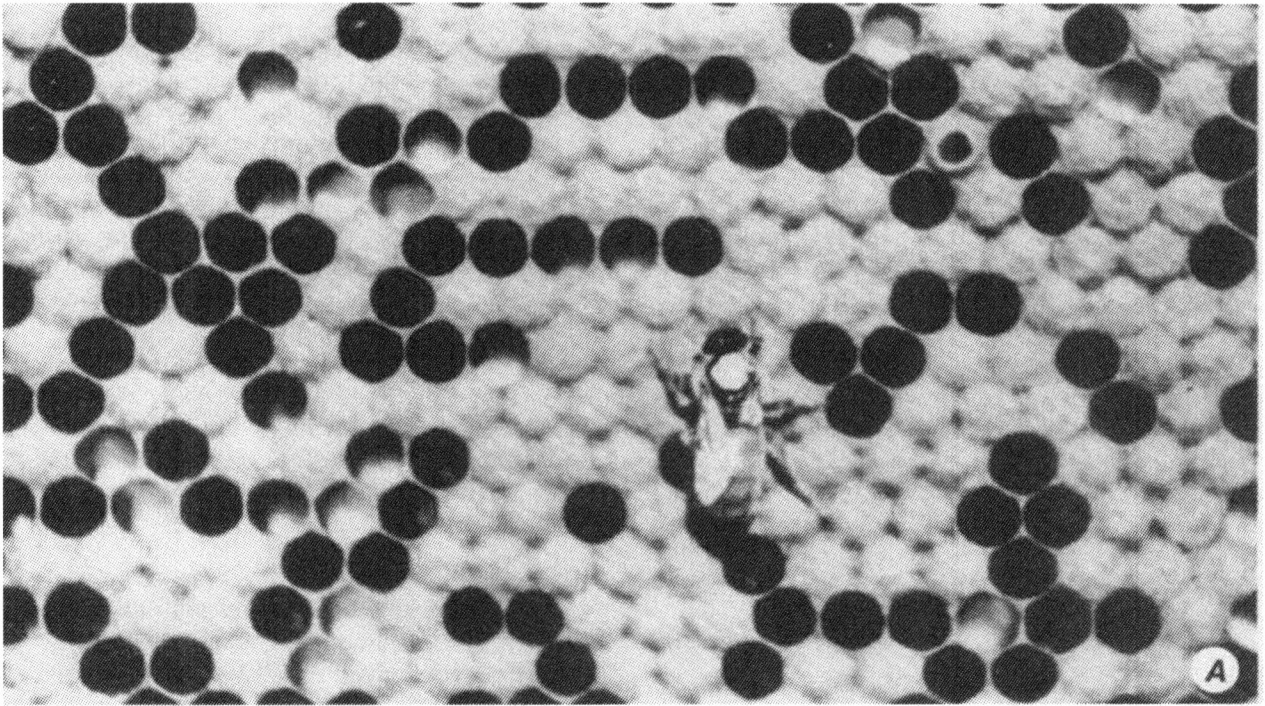

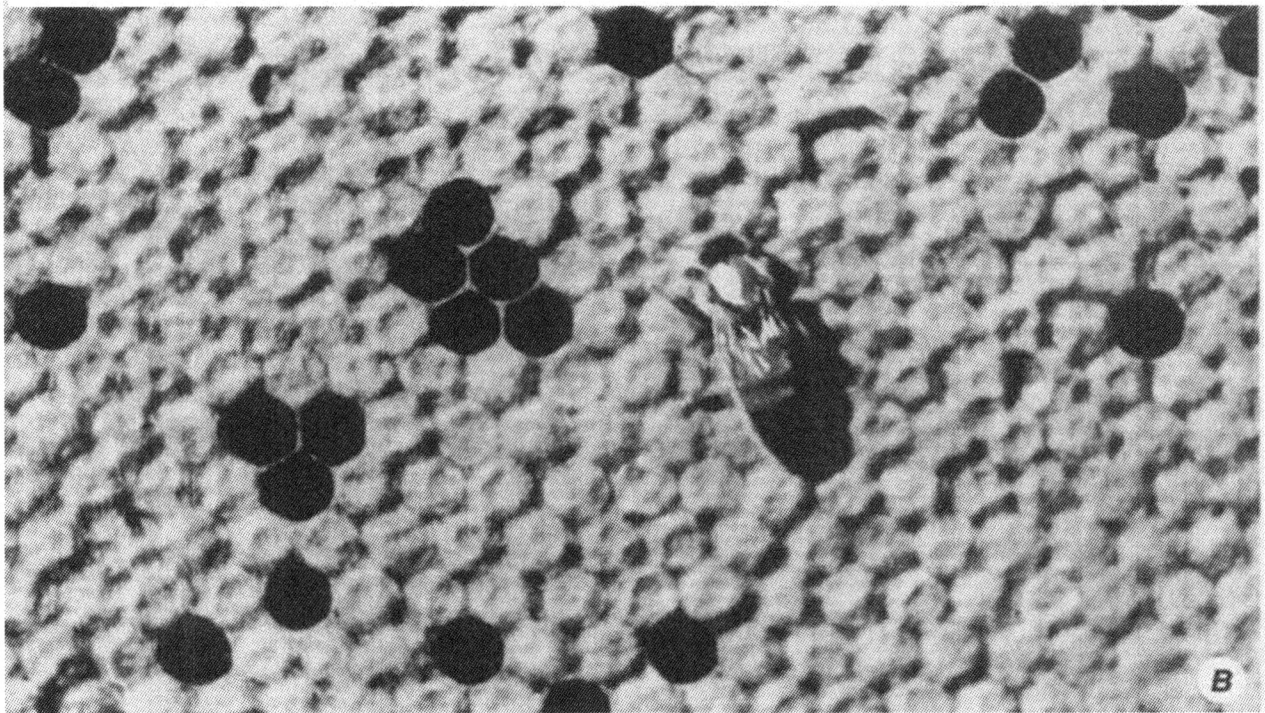

FIGURE 1.—Queens with (A) irregular and (B) good brood pattern.

acceptance is usually obtained when some nectar is available in the field.

One possible period for requeening is during the broodless period of late fall. Queens are easily introduced at this time, and the bees are passive to their presence. However, the uncertainty of the weather, the difficulty of finding old and shrunken queens, and the danger of inciting robbing make this time of year less desirable for requeening than the summer.

Brood rearing declines in late summer and fall, and many normal colonies are completely broodless during much of November and December, particularly if the colony has no pollen. Older queens stop brood rearing sooner than younger queens.

Brood rearing should be encouraged as late in the season as possible. This can be assured by providing vigorous young queens in late summer, by preventing undue overcrowding and restriction of the brood nest with honey, and by encouraging pollen storage.

In areas where fall honey flows occur, partially filled supers should be kept on the colonies, especially if the brood nest is heavy.

If brood rearing is restricted by a crowded brood nest or because of poor queens, the colony may enter the winter with a high percentage of old bees that will die early in the winter. Such colonies may later develop serious nosema infections and perish before spring. A colony should start the winter with about 10 pounds of bees and plenty of honey to carry it to the next spring.

Beekeepers in certain localities will need to think of winter stores for their colonies as early as the first of August if later honey flows are not dependable or are nonexistent. In October, colonies should have at least 45 pounds of honey in dark combs in the top brood chamber and 20 to 30 pounds of honey in each of two lower hive bodies— a total of at least 90 pounds of honey.

Preparing Colony for Winter

Population

The strength of a colony of bees is relative and difficult to describe. A "strong" colony to one beekeeper might be "weak" to another. Colonies with less than 10 pounds of bees should be united to stronger ones or several weaker ones combined. At between 40° and 50°F, 10 pounds of bees will cover practically all the combs of a three-story standard hive wall to wall and top to bottom.

Naturally, as the temperature drops, the cluster will contract.

The beekeeper must see that at no time is the available space for brood rearing reduced because of overcrowding with honey from the fall flow. A balance must be maintained between crowding the colony to get the brood chambers well filled with honey and adding space to relieve brood-rearing restriction. Partly filled supers kept on colonies in the fall may be necessary. Any subnormal colony should not be overwintered but united with another colony.

A colony may appear to have an adequate fall population, but if the bees are old, it will weaken rapidly as winter advances and may starve to death. Starvation occurs even with abundant honey in the hive because the cluster is too small to cover the honey stores.

Food Reserves

The colony should have a minimum of 500 square inches of comb filled with pollen in the fall. To insure uninterrupted brood rearing in late winter and early spring, the beekeeper may need to supplement these stores. The average colony of bees under intensive management may consume about 60 pounds of honey between the last flow in the fall and the first available food from the field in the spring. A weak colony may consume 20 pounds or less, but the very best colony will consume 80 pounds or more. To insure the survival of the top-quality colony, 90 to 100 pounds of honey should be left on it in the fall. A colony of bees not rearing brood will average about one-eighth pound of honey a day or 4 pounds a month. When brood rearing begins, the consumption of honey is greatly accelerated. Brood rearing should start in midwinter and accelerate as temperatures moderate in late winter and early spring.

When brood rearing is discouraged or curtailed, the colony will consume less winter stores but will emerge in the spring much weaker and with a population of primarily old bees. Such colonies will have difficulty replacing the small amount of honey they used over winter, whereas other colonies that have had normal, unimpeded rearing of brood will soon be able to replace all the honey they consumed over winter plus a substantial surplus.

Organization

To accommodate the best queens in standard Langstroth 10-frame hives, a minimum of 2 hive bodies, preferably three, should be used for year-round management. In the fall, most of the honey should be located in the top hive body. With experience, the beekeeper can soon learn to estimate the weight of hive bodies or frames by lifting them. A frame full of honey should weigh approximately 5 pounds. The top hive body should contain 40 to 45 pounds of honey. This means that all frames in the top hive body will be full of honey except for two or three frames in the center. The second body should contain 25 to 30 pounds of honey and some pollen. The bottom hive body should contain 20 to 30 pounds of honey plus pollen. If in the fall the combs in the top hive body are not filled, the beekeeper should reorganize them and if necessary feed additional sugar syrup so that this top hive body is well filled with stores.

As the winter progresses, the cluster of bees will shift its position upward as the stores are consumed. A colony of bees in a cold climate can starve with abundant honey in the hive if the honey is below the cluster.

With the advent of cold weather, the bees cluster tightly in the interspaces of the combs. Usually there are no bees in the bottom part of the hive near the entrance. For this reason, an entrance cleat or reducer should be used to exclude mice, such as 1-inch auger holds drilled into the hive bodies of the brood nest just below the handholds. In late summer, these auger-hole entrances are closed with corks so that the bees will fill the combs near them. During winter the top auger-hold entrance should be open. This allows the escape of moisture-laden air and affords a flight exit for the bees during warm spells (fig. 2).

Packing the Hive

Many beekeepers in the coldest parts of the country consider that some form of protection around the hive is essential. Others believe that colonies with strong populations and ample stores need no further protection. Factors to consider in deciding whether to pack are the cost of material and labor and any savings in honey or bees. Packing will not replenish colonies deficient in honey, pollen, or bees, replace poor queens, or cure bee diseases. Packed colonies will consume slightly less honey. The difference, however, is negligible. The most important consideration in preparing colonies for winter is a strong population and adequate stores.

When outside temperatures are near freezing, the temperature at the surface of a cluster of bees

FIGURE 2.—Colony during winter. Note auger-hole entrance (arrow).

ranges between 43° and 46°F. As the temperature decreases, the cluster contracts and the bees in the outer insulating shell concentrate to provide an insulating band 1 to 3 inches in depth. Metabolism and activity of the bees in the center of the cluster maintain a desired temperature. This may be around 92° if brood rearing is in progress. The temperature of the area of the hive not occupied by bees will be similar to the external temperature. The difference is that the temperature in the unpacked hive changes more rapidly and responds more quickly to that outside the hive. Heavy packing is worse than no packing, because during warm periods in midwinter when the bees should fly, those heavily packed do not fly at all.

It is important to consider the strength of the colony so that the bees can, at all times, cover a good percentage of their winter stores. If the population becomes weakened so that they cannot cover more than a few pounds of honey at a time, they can starve to death because they do not have contact with sufficient food.

Late Winter Manipulation

If colonies are inspected in later winter or early spring, adjustments can be made to save colonies that might be lost otherwise. Even weak or medium-strength colonies often can be saved if honey is moved into contact with the cluster. A strong colony with insufficient honey can starve if additional food is not provided at this time.

From this period until the bees can forage, such colonies can be fed either full combs of honey, or if these are not available, a gallon or two of heavy sugar syrup (two parts sugar by volume to one part water) can be poured directly into the open cells of empty combs.

Spring Buildup

Overwintered colonies usually will start brood rearing in midwinter and continue into the summer unless the stored pollen is all consumed before fresh pollen is available. If the supply is exhausted and not supplemented, brood rearing will slow down or stop entirely when it should proceed without interruption.

For best results in honey production, a beekeeper should have strong populations of young bees for the honey flow. Colonies emerging in the spring with predominantly old bees must build a population of young bees for later flows by using the early sources of pollen.

Some beekeepers trap pollen at the hive entrance from incoming bees by means of a pollen trap such as that described in "Trapping Pollen From Honey Bee Colonies" (Detroy 1976). This pollen is dried or frozen until needed, then mixed with sugar, water, and soy flour, and fed to the colony as a supplement to its natural supply (fig. 3). Various other types of pollen supplements and substitutes

FIGURE 3.—Strong colony feeding on pollen supplement cake.

have been described and some are available on the open market.

Supplements containing pollen are eaten more readily by bees and generally give better results than those containing soy flour or other material without pollen. Pollen supplement is preferred by the bees in direct proportion to the amount of pollen it contains. The less pollen the supplement contains, the less is eaten. Substitutes made without pollen tend to be dry and gummy. A pound of pollen will make approximately 12 pounds of pollen supplement.

Swarm Control in Single-Queen Management

After pollen becomes abundantly available in the spring, the beekeeper should provide ample space for brood rearing and honey storage.

The natural colony behavior is to expand its brood nest upward, and a simple manipulation utilizing this tendency is to shift the empty frames or emerging brood to the top of the hive and the youngest brood and honey to the bottom part. This permits the expansion of the brood rearing upward into this area (fig. 4). Subsequent reversal of brood chambers can be made at about 10-day or 2-week intervals until the honey flow starts.

As soon as the three brood chambers are filled with bees, the first super should be given whether or not the honey flow is in progress. If this is done, most colonies with a vigorous queen will not swarm. However, any queen cells the beekeeper sees as he reverses the brood chambers should be removed. A simple method of reversing brood chambers is to lower the hive backward to the ground, separate the brood chambers, interchange the first and third hive bodies, and return to position.

After the honey flow starts, the danger of swarming lessens and brood chamber reversal can be discontinued. At the start of the honey flow, "bottom supering" should be used. The empty super should be placed above the top brood chamber but below the partially filled supers (fig. 4).

After the supers are filled and the honey extracted, they should never be put directly over the brood nest, but should be placed on top of the partly filled supers to prevent the queen seeking them and laying eggs in them. Why such combs attract the queen is not known.

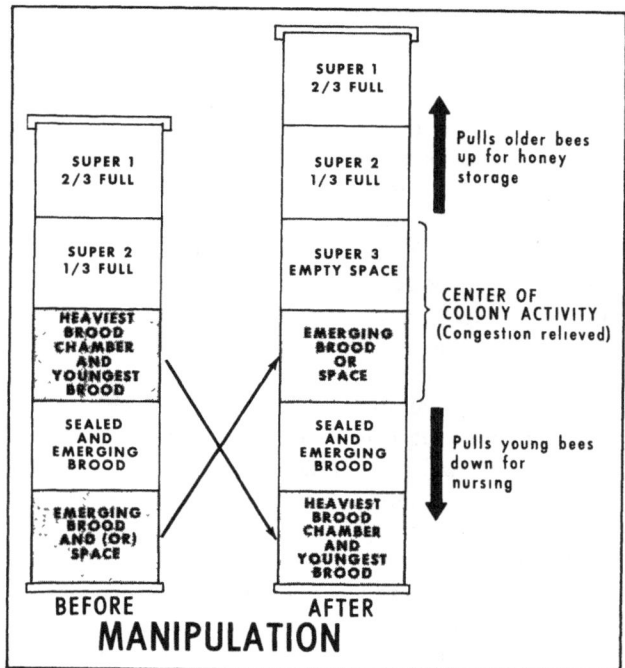

BN-30049

FIGURE 4.—Basic colony manipulation for swarm control.

Two-Queen System

The establishment of a two-queen colony is based on the harmonious existence of two queens in a colony unit. Any system that ensures egg production of two queens in a single colony for about 2 months before the honey flow will boost honey production (Moeller 1956).

The population in a two-queen colony may be twice the population of a single-queen colony. Such a colony will produce more honey and produce it more efficiently than will two single-queen colonies. A two-queen colony usually enters winter with more pollen than a single-queen colony. As a result of this pollen reserve, the two-queen colony emerges in the spring with a larger population of young bees and is thus a more ideal unit for starting another two-queen system.

To operate two-queen colonies, start with strong overwintered colonies. Build them to maximum strength in early spring. Obtain young queens about 2 months before the major honey flows start. When the queens arrive, temporarily divide the colony. Replace the old queen, most of the younger brood, and about half the population in the bottom section. Cover with an inner cover or a thin board and close the escape hole. The division containing most of the sealed and emerging brood, the new queen, and the rest of the population is placed above. The upper unit is provided with an exit hole for flight.

At least two brood chambers must be used for the bottom queen and two for the top queen. Two weeks after the new queen's introduction, remove the division board and replace it with a queen excluder. The supering is double that required for a single-queen operation, or where three standard supers are needed for a single colony, six will be needed for a two-queen colony.

When supering is required, larger populations in two-queen colonies require considerably more room at one time than is required for single-queen colonies. If a single-queen colony receives one super, a two-queen system may require two or even three empty supers at one time.

The brood chambers should be reversed to allow normal upward expansion of the brood area about every 7 to 10 days until about 4 weeks before the expected end of the flow, after which the honey crop on the colony may be so heavy as to preclude any brood nest manipulations. Thereafter, give supers as they are needed for storage of the crop. As the honey is extracted, the supers are returned to the hive to be refilled. They should never be replaced directly over the top brood nest, unless a second queen excluder is used to keep the queen out of them. The top brood nest may tend to become honey bound. If this occurs, reverse the upper and lower brood nests around the queen excluder. This puts the top honey-bound brood nest on the bottom board and the lighter brood nest with the old queen above the excluder.

There is no advantage in having a second queen when about a month of honey flow remains, because eggs laid from this time on will not develop into foragers before the flow has ended. However, entering the brood nest during the middle of the flow to remove one of the queens is impractical. Uniting back to a single-queen status can be done after the bulk of the honey is removed from the colony. By this time some colonies may have already disposed of one queen. When this happens, simply remove the queen excluder and operate the colony as a single-queen unit

Improved Stock

Production of honey is one major criterion in selecting honey bee stock and breeding for improvement. Superior stock must also be reasonably gentle, not prone to excessive swarming, maintain a large but compact brood nest, and winter well. It should ripen its honey rapidly, seal the cells with white wax, and use a minimum of burr comb. To obtain all the desirable characters in a superior stock, specific inbred lines from many sources must be selected and developed and then recombined into a genetically controlled hybrid. When this is done, hybrid vigor or heterosis usually results (Moeller *1976*).

Queens of common stock reared under favorable conditions and heading well-managed colonies probably will be more productive than poorly reared queens of superior stock. Queens of superior stock reared under favorable conditions will require a higher standard of management than is demanded of common stock. To realize the maximum benefits from improved stock, the beekeeper must provide unrestricted room for brood rearing, ripening of nectar, and storage of honey.

The queen breeder should produce the best queens possible to obtain the maximum benefits from improved stock and the honey producer receiving these queens should manage them in such a way that they can develop their maximum colony populations.

Disease Control as Affected by Good Management

If colonies are operated for highest honey yields, they must be kept in optimum condition (fig. 5). This includes rigid control of all bee diseases. For information about bee diseases, see pages 118 to 128.

References

DETROY, B. F. and E. R. HARP
    1976. TRAPPING POLLEN FROM HONEY BEE COLONIES. 11 p. U.S. Department of Agriculture, Production Research Report 163.

FARRAR, C. L.
    1937. INFLUENCE OF COLONY POPULATIONS ON HONEY PRODUCTION. Journal of Agricultural Research 54:945–954.

————
    1942. NOSEMA DISEASE CONTRIBUTES TO WINTER LOSSES AND QUEEN SUPERSEDURE. *In* Gleanings in Bee Culture 70:660–661, 701.

————
    1944. PRODUCTIVE MANAGEMENT OF HONEY BEE COLONIES IN THE NORTHERN STATES. 20 p. U.S. Department of Agriculture Circular 702.

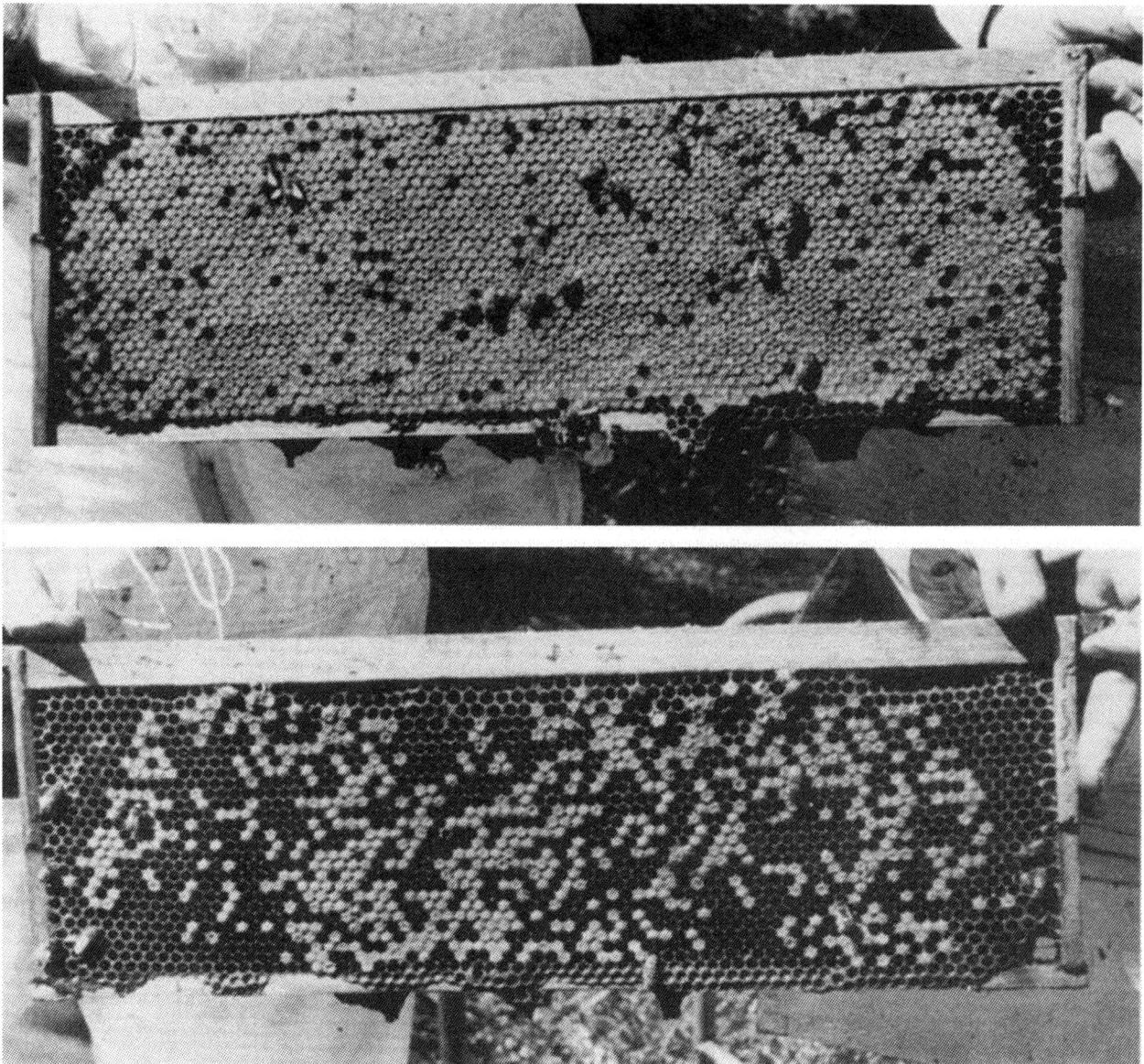

FIGURE 5.—Brood combs showing (top) healthy brood necessary for high honey production and (bottom) diseased brood, which results in weakened colonies and low honey production.

1973–74. PRODUCTIVE MANAGEMENT OF HONEY BEE COLONIES. American Bee Journal 113(8–12), Aug. through Dec. 1973; 114(1–3) Jan. through Mar. 1974.

1952. ECOLOGICAL STUDIES ON OVERWINTERED HONEY BEE COLONIES. Journal of Economic Entomology 45:445–449.

1960. OLD AND NEW IDEAS ABOUT WINTERING. American Bee Journal 100:306–310.

HOOPINGARNER, R., and C. L. FARRAR
  1959. GENETIC CONTROL OF SIZE IN QUEEN HONEY BEES. Journal of Economic Entomology 52:547–548.

MOELLER, F. E.
  1956. BEHAVIOR OF NOSEMA-INFECTED BEES AFFECTING THEIR POSITION IN THE WINTER CLUSTER. Journal of Economic Entomology 49:743–745.

1961. THE RELATIONSHIP BETWEEN COLONY POPULATIONS AND HONEY PRODUCTION AS AFFECTED BY HONEY BEE STOCK LINES. 20 p. U.S. Department of Agriculture, Production Research Report 55.

——— 1976. TWO-QUEEN SYSTEM OF HONEY BEE COLONY MANAGEMENT. 11 p. U.S. Department of Agriculture, Production Research Report 161.

——— 1976. DEVELOPMENT OF HYBRID HONEY BEES. 11 p.

U.S. Department of Agriculture, Production Research Report 168.

SCHAEFER, C. W. and C. L. FARRAR
1946. USE OF POLLEN TRAPS AND POLLEN SUPPLEMENTS IN DEVELOPING HONEY BEE COLONIES. 13 p. U.S. Bureau of Entomology and Plant Quarantine, E–531 revised.

MANAGING COLONIES FOR CROP POLLINATION

By Gordon D. Waller [1]

For centuries honey bee colonies were managed primarily for honey production. Crop pollination occurred as a side benefit to the grower. When fields increased in size and pollination by native pollinators and resident colonies became insufficient, the practice of renting honey bee colonies for crop pollination became commonplace. Today, thousands of colonies are rented by growers of various crops that require insect pollination.

When colonies are used for crop pollination, their value to the grower who rents the bees is measured in fruit or seed produced, not by the honey and wax produced. For the greatest value to the grower the bees must be managed somewhat differently than for honey production. The purpose of this chapter is to present the best information available on managing bees for crop pollination.

Selecting Proper Equipment

Hives

Hives that are to be subjected to frequent moves should be in good condition and of uniform size to facilitate loading and tying down. Telescoping outer covers and bottom boards with landing platforms projecting more than 2 inches prevents the hives from being stacked close together.

Vehicles

Trucks and trailers used to move bees should have smooth flatbeds with hooks around the edges to which ropes can be tied. Pickup trucks and trucks with stake beds make loading and tying down somewhat difficult.

Many beekeepers equip their trucks with a mechanical loader or keep their colonies on pallets and use forklifts for loading to reduce the amount of handlifting. Others use handtrucks (sometimes motorized) and either a ramp or a power-lift tailgate.

Making an Agreement

Contracts

Both the grower and the beekeeper should know precisely what their responsibilities are in relation to the pollination service. A well-understood written agreement should be signed by both parties before the bees are moved to pollinate the crop. The agreement might include the following information: identification of participants, number of colonies, colony strength, rental price, date of delivery to the crop, location or placement of colonies, duration of stay, provisions for water, protection from pesticides, access by the beekeeper, date of removal from the crop, and method of payment, penalties, and rewards.[2]

Colony Evaluation

When a poor set of seed or fruit is attributed to poor pollination, the blame often is put on the beekeeper. Beekeepers cannot guarantee that their bees will work the target crop, but they can guarantee a certain minimum standard colony.

Shortly after the bees have been delivered, the beekeeper and grower should examine a random lot of the colonies to ascertain their quality. Most farmers are not experts in evaluating bee colonies, but they can determine that most of the hive bodies have frames of drawn comb covered with bees, which is better than no information. They also might enlist the aid of a third party who can assess colony strength in terms of number of frames covered with bees or number of frames of brood. The grower should not pay for colonies that are queenless, diseased, or below a minimum population level.

---

[1] Entomologist, Science and Education Administration, Carl Hayden Center for Bee Research, Tucson, Ariz. 85719.

[2] For more information on this subject, see "Pollination Agreements and Services in U.S. Department of Agriculture, Agriculture Handbook No. 496, "Insect Pollination of Cultivated Crop Plants" by S. E. McGregor, or chapter 20, "The Use of Bees for Crop Pollination," in *The Hive and the Honey Bee*, Dadant & Sons, Inc. Hamilton, Ill. 1975.

Adequacy of Pollination

Few guidelines exist that describe adequate bee activity for specific crops. For some crops, the pollination needs have been expressed in terms of bees per 100 flowers, bees per square yard, or bees per 100 feet of row. Unfortunately, few farmers take the time to determine the bee visitation with such precision. Even if they did, the variables of time of day and weather conditions might make such counts somewhat undependable. Recommendations usually vary from one-half to five colonies per acre, but the real criterion is how many bees are working the crop, rather than visiting other attractive plants in the area.

With some crops, such as alfalfa, experienced individuals can assess effectiveness of pollination by the general appearance of the crop (fig. 1). Soon after alfalfa flowers are tripped and pollinated, the petals wilt. When the flowers are not tripped, they remain fresh and the field takes on a showy "flower garden" appearance that contrasts with the dull brownish appearance of a field that has been well-pollinated. Examination of individual florets on several racemes also will reveal the percentage of tripped flowers.

Unfortunately, flowers of many seed crops do not change noticeably following pollination. Only when the crop matures and harvesttime nears can the grower readily judge adequacy of pollination. This is too late to permit remedial changes in pollination to increase the yield. What is needed is a simple, inexpensive field technique by which the plants can be examined periodically to determine if they are being adequately pollinated.

Preparing the Colonies

Stock

Colonies intended for pollination rentals should be headed by young queens of gentle stock to reduce colony failures and to minimize the likelihood of people being stung.

Colony Population

The key to management of colonies for pollination is population control through proper timing of colony development. A colony with a small

FIGURE 1.—Colonies in alfalfa in the West.

population sends forth few foragers. A high percentage of the bees in such a colony must remain in the hive to care for the brood and provide temperature and humidity control. With large populations, a greater proportion of the bees is likely to be involved in collection of nectar and pollen. Since pollination is the result of foraging activity, the colonies should be managed to provide a large number of foraging bees. When beekeepers manage bees primarily for honey production, highly populous colonies (60,000 workers) serve their purpose best. Such large colonies can be extremely difficult to move if four or five stories are required to contain them. Colonies rented for crop pollination are easier to manage if the populations are equal and the hive size is standardized. If the colony has a young queen and adequate comb space for brood and incoming nectar and pollen, it should require a minimum of attention.

To provide adequate colonies in early spring for fruit pollination, the beekeeper must begin preparations the previous fall. Only colonies going into winter with large populations of workers, a young queen, and adequate stores of honey and pollen will be dependable for early spring pollination activity. Conversely, late spring and summer crops, such as cucumbers or alfalfa seed, may require that colonies be divided or otherwise reduced in strength in the spring to prevent them from becoming overcrowded or subject to excessive swarming, or both. A swarm leaves only a remnant of the worker population in the parent hive, which seriously impairs the pollination effectiveness of a colony.

Reserves of Food

Colonies to be moved to a new area for pollination purposes should have ample honey stores. Sometimes, two frames of honey per colony are adequate. At other times, nectar is likely to be scarce and considerably more may be needed. Whatever the amount needed, the beekeeper should see that it is present.

Selecting the New Location

When possible, the beekeeper and farmer should visit the field together during daylight hours to select locations for the placement of colonies. Even when locations are selected in daylight, they often are difficult to recognize at night.

Proximity to the Crop

Farmers who intend to rent bees to pollinate small fields should plan locations that can accommodate groups of colonies nearby. If colonies are to be dispersed throughout larger fields, the distribution, location, and accessibility of colonies should be taken into account. Honey bees can travel several miles to forage, but the likelihood of their visiting a certain crop is decreased as the distance to the crop is increased.

The preferred distribution is to place the colonies in 2 or more groups around the edges of fields less than 20 acres and to distribute them in groups of 6 to 12 about 0.1-mile apart both around and within larger fields (fig. 2).

FIGURE 2.--Palletized colonies placed in groups around the edge of alfalfa seed field for pollination.

Where bloom on other plants is likely to attract foragers away from the target crop, the colonies might be placed on the opposite side of the field from the competition. Neither wind direction nor the direction the entrances face has any effect on where bees will forage—they tend to distribute themselves more or less uniformly in all directions from their hives, if the forage is similar.

Accessibility and Convenience

Proper management requires easy access to the locations to facilitate examination of the colonies, removing honey, or providing additional room for colony expansion. Roadways should be wide enough and turning areas adequate to accommodate the beekeeper's vehicles.

Shade and Water

A nearby source of water is essential for the well-being of a honey bee colony. When water is not available naturally, either the farmer or the beekeeper should provide a temporary source of water for the duration of the pollination job. Large numbers of bees collecting water from stock tanks or around homes may create problems and should be avoided.

Colonies exposed to full sunlight during hot weather expend much energy carrying water to cool their hives. Placing colonies in the shade of trees, under temporary shades, or at least on or near some vegetation increases their efficiency as pollinators.

Protecting the Colonies

One reason beekeepers are hesistant to enter into pollination contracts is the danger of bee kill from pesticides. Even if the crop to be pollinated is not treated with pesticides, nearby crops may be treated and a loss may occur.

If beekeepers are notified a couple of days prior to the time of application of a pesticide, they may be able to protect their colonies by moving them away temporarily or by confining the bees in the hive. Of course, this increases the cost of providing the pollination service.

Consideration also should be given to the likelihood of losses due to vandalism. Colonies placed within full view of an occupied house, and/or located so entry to the area requires passing through a farmyard, are less attractive to vandals than colonies more easily accessible. The likelihood of floods, fires, and wind damage also should be considered when selecting a location.

Colonies placed along busy roadways may have a considerable number of bees killed by passing vehicles. Further, bees entering vehicles are a hazard to motorists. Heavily traveled, unpaved roads create a dust problem for colonies placed nearby—reducing brood viability and colony productivity. For these reasons, when practical, the bees should not be placed adjacent to roads or highways.

Moving Colonies

Timing

The timing of the move to coincide with initiation of bloom on the target crop may be quite important. If moved too early, the bees may forage among other plants and fail to shift to the target crop. If moved too late, the crop yield may be reduced by inadequate pollination of the early flowers. Contact between grower and beekeeper should be established several weeks before the bees are needed. Keep informed on the stage of flowering of the crop so the bees can be provided when needed.

Loading

Most beekeepers prefer to load their colonies into the vehicle in the evening just before dark. At this time, most of the foragers are inside the hive. In the southwest, where daytime temperatures exceed 100°F, flight activity during mid-day is greatly reduced. When colonies are moved at this time, few foragers are lost. Sometimes a weak colony is left behind for stragglers that may be salvaged later.

Confining the Bees

Bees can be kept inside their hives by screening the entrance. Sometimes a screen is placed over the top after the cover has been removed. Closing the entrance in a way that does not allow air circulation might kill the colony, because excessive buildup of heat and carbon dioxide can be lethal. An alternate method of confining the bees is to cover the entire load with a bee-tight plastic or nylon screen. Beekeepers sometimes move colonies with entrances open, if the colonies are to be unloaded before dawn and no stops by the vehicle are contemplated. Beekeepers should make sure that no problems are caused by escaping bees, or bothersome regulations are likely to develop.

The Move

Whenever possible, bees should be moved to their new location immediately after being loaded. In hot weather or on long hauls, spraying the bees with water helps to cool them and prevents dehydration. Immature brood (eggs and larvae) can suffer serious damage from excessively hot and dry conditions. A supply of clean water, readily available at the end of the move, will lessen loss of brood and reduce the chances of the bees collecting water poisoned by earlier insecticide applications.

Managing the Colonies

Feeding

The need for supplemental food will depend on the weather, the crop to be pollinated, and whether other nectar and pollen sources are available to the bees during the pollination period. Colonies used on crops that produce little or no nectar should have adequate stores, as previously mentioned. If short of stores, they can be given sugar syrup or frames of honey. Feeding will have little effect on their nectar foraging activity, and it may help maintain a higher level of brood rearing that ultimately results in greater pollen collection.

Providing Adequate Space

Colonies need adequate room for storage of incoming nectar and pollen, for an expanding brood nest, and for the additional bees that emerge. When not provided adequate space and they become overcrowded, colonies cease foraging if no storage room is available.

Pollination in Cages and Greenhouses

Bees are sometimes used in confined situations such as greenhouses and field cages. The amount of forage available in a cage or greenhouse will not support the bees of a colony, so they should be replaced periodically. If colonies are not provided pollen or pollen supplement, brood rearing soon ceases. If they are not provided with honey or sugar syrup, they will starve.

As a result of these problems, many beekeepers look upon colonies used to pollinate in confined areas as expendable. Some queenless units are provided with no hope for a surviving colony. Small colonies properly managed can survive nicely while providing pollination to caged or greenhouse-grown crops. The question that needs to be answered is one of economics—what should the beekeeper charge, based on the time required to service the colony and on its value at the end of the pollination period?

MOVING COLONIES

By B. F. Detroy [1]

Bees are moved so that they will be near various honey plants or to pollinate orchard and field crops. Some beekeepers also move colonies from northern areas of the United States to the Southern States for overwintering so that colonies may be divided and strong populations will develop early. Sometimes, a honey crop is obtained before the return move north. These moves require equipment for loading the hives to save time and to reduce lifting.

To use labor-saving devices efficiently, certain changes in methods of operation and types of equipment may be needed. The hives should be easily accessible by truck and arranged so that several can be manipulated as a unit. Some type of hive hoist or lift is needed so that the beekeeper can handle heavy colonies alone. Many mechanical devices can be used in loading and unloading colonies. The choice of the appropriate one will depend on the number of colonies to be moved and the frequency and distance of the moves.

The least-expensive lifting device is a crane or hand winch and boom attached to the rear corner of a truck. Many types are built for industrial use. Some can be converted to floor cranes for use in the shop or honey house. These cranes have a reach of about 4 feet, and their length of reach is not changeable. Where they are used, a handcart or similar device is needed to take the hive to them and to move it after it is on the truck. With the lift, one person can place a hive on or off the truck. For the small beekeeper with infrequent moves, this may be satisfactory.

A hydraulic tailgate will lift several hives at one time. Tailgates are made to fit all sizes of trucks. The tailgate is useful for loading hives and supers of honey (fig. 1) and also serves as a platform on which to stand when working with tall hives. The hydraulic system is powered by belt from the truck

engine or power takeoff from the transmission.

Many migratory beekeepers now use either forklifts or hoists. They use cleats on the ends of the hive, and they nail the bottom board underneath. Hives usually are loaded at night, hauled to the new location, and unloaded before dawn. If the hives are loaded with open entrances, a plastic screen covering the entire load should be used.

Boom Loaders

Boom loaders are used in all parts of the country and with a wide range of truck sizes and truck-trailer combinations. With them, a hive can be picked up and placed in position on the truck. The boom loader is mounted on the truck frame, usually between the cab and the truck bed. Two types of drives are used, electrical and hydraulic. An electrically powered hoist operates from the truck battery or from a gasoline-driven generator. Today, most hoists are battery-powered. To keep the battery charged, the truck motor is kept running while the boom is being operated. Battery-operated hoists usually are more dependable than those powered by gasoline engines. The hydraulic hoist operates like the tailgate, but it has electrical valves to control the movement of the boom.

The boom loaders have boom lengths of 12 to 22 feet. The simplest hoist only raises the load, and the operator moves it by hand. Others level automatically and can move the load along the boom. The choice depends largely on how much it will be used. A good hoist should support a 300-pound load.

A fork on the end of the boom is used to pick up the hive. The following three methods are employed: Fork prongs are inserted under the bottom board, fingers are dropped into the hand holes, or fingers are inserted under a 1½-inch cleat nailed on each end of the hive. The cleats usually are the most satisfactory. Hives may be handled singly or two at a time stacked vertically.

---

[1] Agricultural engineer, Science and Education Administration, Bee Management and Entomology Research, University of Wisconsin, Madison, Wis. 53706.

FIGURE 1.—Loading supers on truck with hydraulic tailgate.

On special order, a boom loader can be obtained that will lift a pallet bearing two or four hives with a total maximum weight of 1,000 pounds (fig. 2). A short truck and trailer can be used if the hoist is placed on the rear of the truck or the front of the trailer where it can service both (fig. 3). Booms are difficult to use when colonies are located near trees and cannot be used when colonies are to be placed within an orchard.

The hoist also can be used to lift off the supers for inspection of the colony brood nest. One commercial cart for hand-moving hives is designed to lift the supers and tip them back out of the way, so the brood nest can be inspected. There also is a stationary unit that clamps to the hive, then lifts and swings the supers aside so the brood nest can be inspected.

The most common machine used to load hives on pallets is the four-wheel-drive, skid-steer tractor

FIGURE 2.—Boom loader for lifting pallets containing two or four hives.

Towed Loaders

Generally, where wheel-type loaders are used, they are towed behind the truck on a trailer. A loader that is built on a 4-wheel-drive vehicle chassis usually is towed without a trailer.

The most commonly used trailer is the tandem-wheel type built with little ground clearance. These have hinged loading ramps that can be folded up for towing (fig. 5). Other trailer types have two wheels and are built much heavier than the tandem-wheel type. The wheels are quite large, and the loader is carried much higher above the ground. Hinged ramps also are used for loading. Before a trailer is towed behind the truck, regulations should be checked to see that the combined length does not exceed the permissible limit allowed by the State in which the rig is being operated.

Trucks used for hauling colonies range from the ¾-ton flatbed to the semitractor trailer with a 4-wheel trailer (fig. 6). Most trucks are the straight-flatbed type with beds 16- to 20-feet long and sometimes pull flatbed trailers behind them. Trailers used may be 2-wheel, tandem-wheel, or 4-wheel type. A versatile combination is a small flatbed truck and a 5-wheel type trailer. The bed is removed from the truck when it is used for pulling the trailer. The combination can be used for hauling colonies, and the small truck can be used for other needs.

Beekeepers moving between Northern and Southern States frequently use semitractor-trailer

FIGURE 3.—Trailer-mounted boom loader used to load both truck and trailer.

loader with 1,500-pound lifting capacity (fig. 4). Forklifts mounted on tractors also are frequently used. Some beekeepers build forklift units from four-wheel-drive vehicles. Their advantage over regular forklifts is higher ground speed. Some tractors on which forklifts are mounted have the driver controls reversed so that the operator can observe the loading operation better. Some forklift units have either one or both of the loader forks equipped so that side movement can be hydraulically controlled from the operator's seat to aid in alignment of the loader with the pallets. The tractor also can haul hives to places difficult to reach by truck. If the tractor is equipped with a blade, it can level the area before the hives are unloaded.

FIGURE 4.—A 4-wheel drive, skid-steer tractor with endloader-type loader.

FIGURE 5.—Tandem-wheel trailer towed behind a truck used for hauling skid-steer tractor loaders.

FIGURE 6.—Semitractor trailer with 4-wheel trailer. This unit can carry 468 colonies of honey bees.

FIGURE 7.—A pallet containing seven colonies of honey bees facing in three directions.

combinations. The trailer may be either the flatbed type or enclosed. Enclosed trailers often serve as on-site storage space. Commercial haulers sometimes are employed to haul colonies, but the beekeeper is responsible for loading and unloading the trucks.

Colonies on pallets can be loaded and unloaded from trucks much faster than can single colonies. Usually, four or six colonies are placed back to back on the pallet, with their entrances facing the outside. Some four-colony-per-pallet arrangements have the colonies all facing one direction in pairs of two. Other arrangements are a three-colony pallet, with the colonies in line and all entrances in the same direction, and a seven-colony-per-pallet, with colonies facing out in three directions (fig. 7).

Location of the entrances, especially with the large pallets, can cause management problems. Some six-colony pallets have been replaced with four-colony pallets because the center colonies of six-colony pallets could not be worked.

Some beekeepers use the top of the pallet as the bottom board. Cleats are nailed on the pallet, and a piece of channel-shaped sheet metal is fastened to the pallet to hold the bottom hive body in place. Other pallets are made so that the hives with bottom boards are placed on and held by metal or wood strips or simply strapped to the pallet. In humid areas, the hives are spaced 1 to 2 inches apart on the pallet to reduce wood rot.

Some beekeepers have designed pallets the width of the truck. These are pulled on or off the truck by special drag chains.

References

DETROY, B. F., C. D. OWENS, and L. O. WHITEFOOT.
  1975. MOVING COLONIES OF HONEY BEES. American Bee Journal 115:268–271.
OWENS, C. D., and B. F. DETROY.
  1977. SELECTING AND OPERATING BEEKEEPING EQUIPMENT. 24 p. U.S. Department of Agriculture, Farmers' Bulletin 2204, revised.

HONEY COMPOSITION AND PROPERTIES

By J. W. WHITE, JR. AND LANDIS W. DONER [1]

Honey is essentially a highly concentrated water solution of two sugars, dextrose and levulose, with small amounts of at least 22 other more complex sugars. Many other substances also occur in honey, but the sugars are by far the major components. The principal physical characteristics and behavior of honey are due to its sugars, but the minor constituents—such as flavoring materials, pigments, acids, and minerals—are largely responsible for the differences among individual honey types.

Honey, as it is found in the hive, is a truly remarkable material, elaborated by bees with floral nectar, and less often with honeydew. Nectar is a thin, easily spoiled sweet liquid that is changed ("ripened") by the honey bee to a stable, high-density, high-energy food. The earlier U.S. Food and Drug Act defined honey as "the nectar and saccharine exudation of plants, gathered, modified, and stored in the comb by honey bees (*Apis melli-fera* and *A. dorsata*); is levorotatory; contains not more than 25% water, not more than 0.25% ash, and not more than 8% sucrose." The limits established in this definition were largely based on a survey published in 1908. Today, this definition has an advisory status only, but is not totally correct, as it allows too high a content of water and sucrose, is too low in ash, and makes no mention of honeydew.

Colors of honey form a continuous range from very pale yellow through ambers to a darkish red amber to nearly black. The variations are almost entirely due to the plant source of the honey, although climate may modify the color somewhat through the darkening action of heat.

The flavor and aroma of honey vary even more than the color. Although there seems to be a characteristic "honey flavor," almost an infinite number of aroma and flavor variations can exist.

As with color, the variations appear to be governed by the floral source. In general, light-colored honey is mild in flavor and a darker honey has a more pronounced flavor. Exceptions to the rule sometimes endow a light honey with very definite specific flavors. Since flavor and aroma judgments are personal, individual preference will vary, but with the tremendous variety available, everyone should be able to find a favorite honey.

Composition of Honey

By far, the largest portion of the dry matter in honey consists of the sugars. This very concentrated solution of several sugars results in the characteristic physical properties of honey—high viscosity, "stickiness," high density, granulation tendencies, tendency to absorb moisture from the air, and immunity from some types of spoilage. Because of its unique character and its considerable difference from other sweeteners, chemists have long been interested in its composition and food technologists sometimes have been frustrated in attempts to include honey in prepared food formulas or products. Limitations of methods available to earlier researchers made their results only approximate in regard to the true sugar composition of honey. Although recent research has greatly improved analytical procedures for sugars, even now some compromises are required to make possible accurate analysis of large numbers of honey samples for sugars.

An analytical survey of U.S. honey is reported in *Composition of American Honeys*, Technical Bulletin 1261, published by the U.S. Department of Agriculture in 1962. In this survey, considerable effort was made to obtain honey samples from all over the United States and to include enough samples of the commercially significant floral types that the results, averaged by floral type, would be useful to the beekeeper and packer and also to the food technologist. In addition to providing tables of composition of U.S. honeys, some

---

[1] Research leader and research chemist, respectively, Science and Education Administration, Eastern Regional Research Center, Philadelphia, Pa. 19118.

general conclusions were reached in the bulletin on various factors affected by honey composition.

Where comparisons were made of the composition of the same types of honey from 2 crop years, relatively small or no differences were found. The same was true for the same type of honey from various locations. As previously known, dark honey is higher than light honey in ash (mineral) and nitrogen content. Averaging results by regions showed that eastern and southern honeys were darker than average, whereas north-central and intermountain honeys were lighter. The north-central honey was higher than average in moisture, and the intermountain honey was more heavy bodied. Honey from the South Atlantic States showed the least tendency to granulate, whereas the intermountain honey had the greatest tendency.

The technical bulletin includes complete analyses of 490 samples of U.S. floral honey and 14 samples of honeydew honey gathered from 47 of the 50 States and representing 82 "single" floral types and 93 blends of "known" composition. For the more common honey types, many samples were available and averages were calculated by computer for many floral types and plant families. Also given in this bulletin are the average honey composition for each State and region and detailed discussions of the effects of crop year, storage, area of production, granulation, and color on composition. Some of the tabular data are included in this handbook.

Table 1 gives the average value for all of the constituents analyzed in the survey and also lists the range of values for each constitutent. The range shows the great variability for all honey constituents. Most of the constituents listed are familiar. Levulose and dextrose are the simple sugars making up most of the honey. Fructose and glucose are other commonly used names for these sugars. Sucrose (table sugar) also is present in honey, and is one of the main sugars in nectar, along with levulose and dextrose. "Maltose" is actually a mixture of several complex sugars, which are analyzed collectively and reported as maltose.

TABLE 1.—*Average composition of floral and honeydew honey and range of values* [1]

Characteristic or constituent	Floral honey		Honeydew honey	
	Average values	Range of values	Average values	Range of values
Color [2]	Dark half of white.	Light half of water white to dark.	Light half of amber.	Dark half of extra light amber to dark.
Granulating tendency [3]	Few clumps of crystals ⅛- to ¼-inch layer.	Liquid to complete hard granulation.	¹⁄₁₆- to ⅛-inch layer of crystals.	Liquid to complete soft granulation.
Moisture_____percent__	17.2	13.4–22.9	16.3	12.2–18.2.
Levulose_____do____	38.19	27.25–44.26	31.80	2 .91–38.12.
Dextrose_____do____	31.28	22.03–40.75	26.08	19.23–31.86.
Sucrose_____do____	1.31	.25–7.57	.80	.44–1.14.
Maltose_____do____	7.31	2.74–15.98	8.80	5.11–12.48.
Higher sugars_____do____	1.50	.13–8.49	4.70	1.28–11.50.
Undetermined_____do____	3.1	0–13.2	10.1	2.7–22.4.
pH	3.91	3.42–6.10	4.45	3.90–4.88.
Free acidity [4]	22.03	6.75–47.19	49.07	30.29–66.02.
Lactone [4]	7.11	0–18.76	5.80	.36–14.09.
Total acidity [4]	29.12	8.68–59.49	54.88	34.62–76.49.
Lactone÷free acid	.335	0–.950	.127	.007–.385
Ash_____percent__	.169	.020–1.028	.736	.212–1.185.
Nitrogen_____do____	.041	0–.133	.100	.047–.223.
Diastase [5]	20.8	2.1–61.2	31.9	6.7–48.4.

[1] Based on 490 samples of floral honey and 14 samples of honeydew honey.

[2] Expressed in terms of U.S. Department of Agriculture color classes.

[3] Extent of granulation for heated sample after 6 months' undisturbed storage.

[4] Milliequivalents per kilogram.

[5] 270 samples for floral honey.

Higher sugars is a more descriptive term for the material formerly called honey dextrin.

The undetermined value is found by adding all the sugar percentages to the moisture value and subtracting from 100. The active acidity of a material is expressed as pH; the larger the number the lower is the active acidity. The lactone is a newly found component of honey. Lactones may be considered to be a reserve acidity, since by chemically adding water to them (hydrolysis) an acid is formed. The ash is, of course, the material remaining after the honey is burned and represents mineral matter. The nitrogen is a measure of the protein material, including the enzymes, and diastase is a specific starch-digesting enzyme.

Most of these constituents are expressed in percent, that is, parts per hundred of honey. The acidity is reported differently. In earlier times, acidity was reported as percent formic acid. We now know that there are many acids in honey, with formic acid being one of the least important. Since a sugar acid, gluconic acid, has been found to be the principal one in honey, these results could be expressed as "percent gluconic acid" by multiplying the numbers in the table by 0.0196. Since actually there are many acids in honey, the term "milliequivalents per kilogram" is used to avoid implying that only one acid is found in honey. This figure is such that it properly expresses the acidity of a honey sample independently of the kind or kinds of acids present.

In table 1, the differences between floral honey and honeydew honey[2] can be seen. Floral honey is higher in simple sugars (levulose and dextrose), lower in disaccharides and higher sugars (dextrins), and contains much less acid. The higher amount of mineral salts (ash) in honeydew gives it a less active acidity (higher pH). The nitrogen content reflecting the amino acids and protein content is also higher in honeydew.

The main sugars in the common types of honey are shown in table 2. Levulose is the major sugar in all the samples, but there are a few types, not on the list, that contain more dextrose than levulose (dandelion and the blue curls). This excess of levulose over dextrose is one way that honey differs from commercial invert sugar. Even though honey has less dextrose than levulose, it is dextrose that crystallizes when honey granu-

[2] Strickly speaking, honeydew is an excretory product of several species of insects that suck plant juices. If it is gathered and stored by bees, it becomes honeydew honey.

lates, because it is less soluble in water than is levulose. Even though honey contains an active sucrose-splitting enzyme, the sucrose level in honey never reaches zero.

Honey varies tremendously in color and flavor, depending largely on its floral source. Its composition also varies widely, depending on its floral sources (table 2). Although hundreds of kinds of honey are produced in this country, only about 25 or 30 are commercially important and available in large quantities. Until the comprehensive survey of honey composition was published in 1962, the degree of compositional variation was not known. This lack of information hindered the widespread use of honey by the food industry.

Water Content

The natural moisture of honey in the comb is that remaining from the nectar after ripening. The amount of moisture is a function of the factors involved in ripening, including weather conditions and original moisture of the nectar. After extraction of the honey, its moisture content may change, depending on conditions of storage. It is one of the most important characteristics of honey influencing keeping quality, granulation, and body.

Beekeepers as well as honey buyers know that the water content of honey varies greatly. It may range between 13 and 25 percent. According to the United States Standards for Grades of Extracted Honey, honey may not contain more than 18.6 percent moisture to qualify for U.S. grade A (U.S. Fancy) and U.S. grade B (U.S. Choice). Grade C (U.S. Standard) honey may contain up to 20 percent water; any higher amount places a honey in U.S. grade D (Substandard).

These values represent limits and do not indicate the preferred or proper moisture content for honey. If honey has more than 17 percent moisture and contains a sufficient number of yeast spores, it will ferment. Such honey should be pasteurized, that is, heated sufficiently to kill such organisms. This is particularly important if the honey is to be "creamed" or granulated, since this process results in a slightly higher moisture level in the liquid part. On the other hand, it is possible for honey to be too low in moisture from some points of view. In the West, honey may have a moisture content as low as 13 to 14 percent. Such honey is somewhat difficult to handle, though it is most useful in blending to reduce moisture content. It

TABLE 2.—*Carbohydrate composition of honey types*

Number of samples	Floral type	Dextrose	Levulose	Sucrose	Maltose	Higher sugars
		Percent	*Percent*	*Percent*	*Percent*	*Percent*
23	Alfalfa	33. 40	39. 11	2. 64	6. 01	. 89
25	Alfalfa-sweetclover	33. 57	39. 29	2. 00	6. 30	. 91
5	Aster	31. 33	37. 55	. 81	8. 45	1. 04
3	Basswood	31. 59	37. 88	1. 20	6. 86	1. 44
3	Blackberry	25. 94	37. 64	1. 27	11. 33	2. 50
5	Buckwheat	29. 46	35. 30	. 78	7. 63	2. 27
4	Buckwheat, wild	30. 50	39. 72	. 79	7. 21	. 83
26	"Clover"	32. 22	37. 84	1. 44	6. 60	1. 39
3	Clover, alsike	30. 72	39. 18	1. 40	7. 46	1. 55
3	Clover, crimson	30. 87	38. 21	. 91	8. 59	1. 63
3	Clover, Hubam	33. 42	38. 69	. 86	6. 23	. 74
10	Cotton	36. 74	39. 28	1. 14	4. 87	. 50
3	Fireweed	30. 72	39. 81	1. 28	7. 12	2. 06
6	Gallberry	30. 15	39. 85	. 72	7. 71	1. 22
3	Goldenrod	33. 15	39. 57	. 51	6. 57	. 59
2	Heartsease	32. 98	37. 23	1. 95	5. 71	. 53
2	Holly	25. 65	38. 98	1. 00	10. 07	2. 16
3	Honeydew, cedar	25. 92	25. 16	. 68	6. 20	9. 61
5	Honeydew, oak	27. 43	34. 84	. 84	10. 45	2. 16
2	Horsemint	33. 63	37. 37	1. 01	5. 53	. 73
3	Locust, black	28. 00	40. 66	1. 01	8. 42	1. 90
3	Loosestrife, purple	29. 90	37. 75	. 62	8. 13	2. 35
3	Mesquite	36. 90	40. 41	. 95	5. 42	. 35
4	Orange, California	32. 01	39. 08	2. 68	6. 26	1. 23
13	Orange, Florida	31. 96	38. 91	2. 60	7. 29	1. 40
4	Raspberry	28. 54	34. 46	. 51	8. 68	3. 58
3	Sage	28. 19	40. 39	1. 13	7. 40	2. 38
3	Sourwood	24. 61	39. 79	. 92	11. 79	2. 44
4	Star-thistle	31. 14	36. 91	2. 27	6. 92	2. 74
8	Sweetclover	30. 97	37. 95	1. 41	7. 75	1. 40
3	Sweetclover, yellow	32. 81	39. 22	2. 94	6. 63	. 97
4	Tulip tree	25. 85	34. 65	. 69	11. 57	2. 96
5	Tupelo	25. 95	43. 27	1. 21	7. 97	1. 11
7	Vetch	31. 67	38. 33	1. 34	7. 23	1. 83
9	Vetch, hairy	30. 64	38. 20	2. 03	7. 81	2 08
12	White clover	30. 71	38. 36	1. 03	7. 32	1. 56

contains over 6 percent more honey solids than a product of 18.6 percent moisture.

In the 490 samples of honey analyzed in the Department's Technical Bulletin 1261, the average moisture content was 17.2 percent. Samples ranged between 13.4 and 22.9 percent, and the standard deviation was 1.46. This means that 68 percent of the samples (or of all U.S. honey) will fall within the limits of 17.2 ± 1.46 percent moisture (15.7–18.7); 95.5 percent of all U.S. honey will fall within the limits of 17.2 ± 2.92 percent moisture (14.3–20.1).

In the same bulletin, a breakdown of average moisture contents by geographic regions is shown. These values (percent) are North Atlantic, 17.3; East North Central, 18.0; West North Central, 18.2; South Atlantic, 17.7; South Central, 17.5; Intermountain West, 16.0; and West, 16.1.

Sugars

Honey is above all a carbohydrate material, with 95 to 99.9 percent of the solids being sugars, and the identity of these sugars has been studied for many years. Sugars are classified according to their size or the complexity of the molecules of which they are made. Dextrose (glucose) and levulose (fructose), the main sugars in honey, are simple sugars, or monosaccharides, and are the building blocks for the more complex honey sugars. Dextrose and levulose account for about 85 percent of the solids in honey.

Until the middle of this century, the sugars of honey were thought to be a simple mixture of dextrose, levulose, sucrose (table sugar), and an ill-defined carbohydrate material called "honey dextrin." With the advent of new methods for separating and analyzing sugars, workers in Europe, the United States, and Japan have identified many sugars in honey after separating them from the complex honey mixture. This task has been accomplished using a variety of physical and chemical methods.

Dextrose and levulose are still by far the major sugars in honey, but 22 others have been found. All of these sugars are more complex than the monosaccharides, dextrose and levulose. Ten disaccharides have been identified: sucrose, maltose, isomaltose, maltulose, nigerose, turanose, kojibiose, laminaribiose, α, β-trehalose, and gentiobiose. Ten trisaccharides are present: melezitose, 3-α-isomaltosylglucose, maltotriose, l-kestose, panose, isomaltotriose, erlose, theanderose, centose, and isopanose. Two more complex sugars, isomaltotetraose and isomaltopentaose, have been identified. Most of these sugars are present in quite small quantities.

Most of these sugars do not occur in nectar, but are formed either as a result of enzymes added by the honeybee during the ripening of honey or by chemical action in the concentrated, somewhat acid sugar mixture we know as honey.

Acids

The flavor of honey results from the blending of many "notes," not the least being a slight tartness or acidity. The acids of honey account for less than 0.5 percent of the solids, but this level contributes not only to the flavor, but is in part responsible for the excellent stability of honey against micro-organisms. Several acids have been found in honey, gluconic acid being the major one. It arises from dextrose through the action of an enzyme called glucose oxidase. Other acids in honey are formic, acetic, butyric, lactic, oxalic, succinic, tartaric, maleic, pyruvic, pyroglutamic, α-ketoglutaric, glycollic, citric, malic, 2- or 3-phosphoglyceric acid, α- or β-glycerophosphate, and glucose 6-phosphate.

Proteins and Amino Acids

It will be noted in table 1 that the amount of nitrogen in honey is low, 0.04 percent on the average, though it may range to 0.1 percent. Recent work has shown that only 40 to 65 percent of the total nitrogen in honey is in protein, and some nitrogen resides in substances other than proteins, namely the amino acids. Of the 8 to 11 proteins found in various honeys, 4 are common to all, and appear to originate in the bee, rather than the nectar. Little is known of many proteins in honey, except that the enzymes fall into this class.

The presence of proteins causes honey to have a lower surface tension than it would have otherwise, which produces a marked tendency to foam and form scum and encourages formation of fine air bubbles. Beekeepers familiar with buckwheat honey know how readily it tends to foam and produce surface scum, which is largely due to its relatively high protein content.

The amino acids are simple compounds obtained when proteins are broken down by chemical or digestive processes. They are the "building blocks" of the proteins. Several of them are essential to life and must be obtained in the diet. The quantity of free amino acids in honey is small and of no nutritional significance. Breakthroughs in the separation and analysis of minute quantities of material (chromatography) have revealed that various honeys contain 11 to 21 free amino acids. Proline, glutamic acid, alanine, phenylalanine, tyrosine, leucine, and isoleucine are the most common, with proline predominating.

Amino acids are known to react slowly, or more rapidly by heating, with sugars to produce yellow or brown materials. Part of the darkening of honey with age or heating may be due to this.

Minerals

When honey is dried and burned, a small residue of ash invariably remains, which is the mineral content. As shown in table 1, it varies from 0.02 to slightly over 1 percent for a floral honey, averaging about 0.17 percent for the 490 samples analyzed.

Honeydew honey is richer in minerals, so much so that its mineral content is said to be a prime cause of its unsuitability for winter stores. Schuette and his colleagues at the University of Wisconsin have examined the mineral content of light and dark honey. They reported the following average values:

Mineral	Light honey (p p m.)	Dark honey (p.p.m.)
Potassium	205	1, 676
Chlorine	52	113
Sulfur	58	100
Calcium	49	51
Sodium	18	76
Phosphorus	35	47
Magnesium	19	35
Silica	22	36
Iron	2. 4	9. 4
Manganese	. 30	4. 09
Copper	. 29	. 56

Enzymes

One of the characteristics that sets honey apart from all other sweetening agents is the presence of enzymes. These conceivably arise from the bee, pollen, nectar, or even yeasts or micro-organisms in the honey. Those most prominent are added by the bee during the conversion of nectar to honey. Enzymes are complex protein materials that under mild conditions bring about chemical changes, which may be very difficult to accomplish in a chemical laboratory without their aid. The changes that enzymes bring about throughout nature are essential to life.

Some of the most important honey enzymes are invertase, diastase, and glucose oxidase.

Invertase, also known as sucrase or saccharase, splits sucrose into its constituent simple sugars, dextrose, and levulose. Other more complex sugars have been found recently to form in small amounts during this action and in part explain the complexity of the minor sugars of honey. Although the work of invertase is completed when honey is ripened, the enzyme remains in the honey and retains its activity for some time. Even so, the sucrose content of honey never reaches zero. Since the enzyme also synthesizes sucrose, perhaps the final low value for the sucrose content of honey represents an equilibrium between splitting and forming sucrose.

Diastase (amylase) digests starch to simpler compounds but no starch is found in nectar. What its function is in honey is not clear. Diastase appears to be present in varying amounts in nearly all honey and it can be measured. It has probably had the greatest attention in the past, because it has been used as a measure of honey quality in several European countries.

Glucose oxidase converts dextrose to a related material, a gulconolactone, which in turn forms gluconic acid, the principal acid in honey. Since this enzyme previously was shown to be in the pharyngeal gland of the honey bee, this is probably the source. Here, as with other enzymes, the amount varies in different honeys. In addition to gluconolactone, glucose oxidase forms hydrogen peroxide during its action on dextrose, which has been shown to be the basis of the heat-sensitive antibacterial activity of honey.

Other enzymes are reported to be present in honey, including catalase and an acid phosphatase. All the honey enzymes can be destroyed or weakened by heat.

Properties of Honey

Because of honey's complex and unusual composition, it has several interesting attributes. In addition, honey has some properties, because of its composition, that make it difficult to handle and use. With modern technology, however, methods have been established to cope with many of these problems.

Antibacterial Activity

An ancient use for honey was in medicine as a dressing for wounds and inflammations. Today, medicinal uses of honey are largely confined to folk medicine. On the other hand, since milk can be a carrier of some diseases, it was once thought that honey might likewise be such a carrier. Some years ago this idea was examined by adding nine common pathogenic bacteria to honey. All the bacteria died within a few hours or days. Honey is not a suitable medium for bacteria for two reasons—it is fairly acid and it is too high in sugar content for growth to occur. This killing of bacteria by high sugar content is called osmotic effect. It seems to function by literally drying out the bacteria. Some bacteria, however, can survive in the resting spore form, though not grown in honey.

Another type of antibacterial property of honey is that due to inhibine. The presence of an antibacterial activity in honey was first reported about 1940 and confirmed in several laboratories. Since then, several papers were published on this subject. Generally, most investigators agree that inhibine (name used by Dold, its discoverer, for antibacterial activity) is sensitive to heat and light. The effect of heat on the inhibine content of honey was studied by several investigators. Apparently, heating honey sufficiently to reduce markedly or to destroy its inhibine activity would deny it a market as first-quality honey in several European countries. The use of sucrase and inhibine assays together was proposed to determine the heating history of commercial honey.

Until 1963, when White showed that the inhibine effect was due to hydrogen peroxide produced and accumulated in diluted honey, its identity remained unknown. This material, well known for its antiseptic properties, is a byproduct of the formation of gluconic acid by an enzyme that occurs in honey, glucose oxidase. The peroxide can inhibit the growth of certain bacteria in the diluted honey. Since it is destroyed by other honey constituents, an equilibrium level of peroxides will occur in a diluted honey, its magnitude depending on many factors such as enzyme activity, oxygen availability, and amounts of peroxide-destroying materials in the honey. The amount of inhibine (peroxide accumulation) in honey depends on floral type, age, and heating.

A chemical assay method has been developed that rapidly measures peroxide accumulation in diluted honey. By this procedure, different honeys have been found to vary widely in the sensitivity of their inhibine to heat. In general, the sensitivity is about the same as or greater than that of invertase and diastase in honey.

Food Value

Honey is primarily a high-energy carbohydrate food. Because its distinct flavors cannot be found elsewhere, it is an enjoyable treat. The honey sugars are largely the easily digestible "simple sugars," similar to those in many fruits. Honey can be regarded as a good food for both infants and adults.

The protein and enzymes of honey, though used as indicators of heating history and hence table quality in some countries, are not present in sufficient quantities to be considered nutritionally significant. Several of the essential vitamins are present in honey, but in insignificant levels. The mineral content of honey is variable, but darker honeys have significant quantities of minerals.

Granulation

Dextrose, a major sugar in honey, can spontaneously crystallize from any honeys in the form of its monohydrate. This sometimes occurs when the moisture level in honey is allowed to drop below a certain level.

A large part of the honey sold to consumers in the United States is in the liquid form, much less in a finely granulated form known as "honey spread" or finely granulated honey, and even less as comb honey. The consumer appears to be conditioned to buying liquid honey. At least sales of the more convenient spread form have never approached those of liquid honey.

Since the granulated state is natural for most of the honey produced in this country, processing is required to keep it liquid. Careful application of heat to dissolve "seed" crystals and avoidance of subsequent "seeding" will usually suffice to keep a honey liquid for 6 months. Damage to color and flavor can result from excessive or improperly applied heat. Honey that has granulated can be returned to liquid by careful heating. Heat should be applied indirectly by hot water or air, not by direct flame or high-temperature electrical heat. Stirring accelerates the dissolution of crystals. For small containers, temperatures of 140°F for 30 minutes usually will suffice.

If unheated honey is allowed to granulate naturally, several difficulties may arise. The texture may be fine and smooth or granular and objectionable to the consumer. Furthermore, a granulated honey becomes more susceptible to spoilage by fermentation, caused by natural yeast found in all honeys and apiaries. Quality damage from poor texture and fermented flavors usually is far greater than any caused by the heat needed to eliminate these problems.

Finely granulated honey may be prepared from a honey of proper moisture content (17.5 percent in summer, 18 percent in winter) by several processes. All involve pasteurization to eliminate fermentation, followed by addition at room temperature of 5 to 10 percent of a finely granulated "starter" of acceptable texture, thorough mixing, and storage at 55° to 60°F in the retail containers for about a week. The texture remains acceptable if storage is below about 80° to 85°.

Deterioration of Quality

Fermentation.—Fermentation of honey is caused by the action of sugar-tolerant yeasts upon the sugars dextrose and levulose, resulting in the formation of ethyl alcohol and carbon dioxide. The alcohol in the presence of oxygen then may be broken down into acetic acid and water. As a result, honey that has fermented may taste sour.

The yeasts responsible for fermentation occur naturally in honey, in that they can germinate and grow at much higher sugar concentrations than other yeasts, and, therefore, are called "osmophilic." Even so there are upper limits of sugar concentration beyond which these yeasts will not grow. Thus, the water content of a honey is one of the factors concerned in spoilage by fermentation. The others are extent of contamination by yeast spores (yeast count) and temperature of storage.

Honey with less than 17.1 percent water will not ferment in a year, irrespective of the yeast count. Between 17.1 and 18 percent moisture, honey with 1,000 yeast spores or less per gram will be safe for a year. When moisture is between 18.1 and 19 percent, not more than 10 yeast spores per gram can be present for safe storage. Above 19 percent water, honey can be expected to ferment even with only one spore per gram of honey, a level so low as to be very rare.

When honey granulates, the resulting increased moisture content of the liquid part is favorable for fermentation. Honey with a high moisture content will not ferment below 50°F or above about 80°. Honey even of relatively low water content will ferment at 60°. Storing at temperatures over 80° to avoid fermentation is not practical as it will damage honey.

E. C. Martin has studied the mechanism and course of yeast fermentation in honey in conjunction with his work on the hygroscopicity of honey. He confirmed that when honey absorbs moisture, which occurs when it is stored above 60-percent relative humidity, the moisture content at first increases mostly at the surface before the water diffuses into the bulk of the honey. When honey absorbs moisture, yeasts grow aerobically (using oxygen) at the surface and multiply rapidly, whereas below the surface the growth is slower.

Fermenting honey is usually at least partly granulated and is characterized by a foam or froth on the surface. It will foam considerably when heated. An odor as of sweet wine or fermenting fruit may be detected. Gas production may be so vigorous as to cause honey to overflow or burst a container. The off-flavors and odors associated with fermentation probably arise from the acids produced by the yeasts.

Honey that has been fermented can sometimes be reclaimed by heating it to 150°F for a short time. This stops the fermentation and expels some of the off-flavor. Fermentation in honey may be avoided by heating to kill yeasts. Minimal treatments to pasteurize honey are as follows:

Temperature (°F):	Heating time (minutes)
128	470
130	170
135	60
140	42
145	7.5
150	2.8
155	1.0
160	.4

The following summarize the important aspects of fermentation:

1. All honey should be considered to contain yeasts.

2. Honey is more liable to fermentation after granulation.

3. Honey of over 17 percent water may ferment and over 19 percent water will ferment.

4. Storage below 50°F will prevent fermentation during such storage, but not later.

5. Heating honey to 150°F for 30 minutes will destroy honey yeasts and thus prevent fermentation.

Quality loss by heating and storing.—The other principal types of honey spoilage, damage by overheating and by improper storing, are related to each other. In general, changes that take place quickly during heating also occur over a longer period during storage with the rate depending on the temperature. These include darkening, loss of fresh flavor, and formation of off-flavor (caramelization).

To keep honey in its original condition of high quality and delectable flavor and fragrance is possibly the greatest responsibility of the beekeeper and honey packer. At the same time it is an operation receiving perhaps less attention from the producer than any other and one requiring careful consideration by packers and wholesalers. To do an effective job, one must know the factors that govern honey quality, as well as the effects of various beekeeping and storage practices on honey quality. The factors are easily determined, but only recently are the facts becoming known regarding the effects of processing temperatures and storage on honey quality.

To be of highest quality, a honey—whether liquid, crystallized, or comb—must be well ripened with proper moisture content; it must be free of extraneous materials, such as excessive pollen, dust, insect parts, wax, and crystals if liquid; it must not ferment; and above all it must be of excellent flavor and aroma, characteristic of the particular honey type. It must, of course, be free of off-flavors or odors of any origin. In fact, the more closely it resembles the well-ripened honey as it exists in the cells of the comb, the better it is.

Several beekeeping practices can reduce the quality of the extracted product. These include combining inferior floral types, either by mixing at extracting time or removing the crop at incorrect times, extraction of unripe honey, extraction of brood combs, and delay in settling and straining. However, we are concerned here with the handling of honey from its extraction to its sale. During this time improper settling, straining, heating, and storage conditions can make a superb honey into just another commercial product.

The primary objective of all processing of honey is simple—to stabilize it. This means to keep it free of fermentation and to keep the desired physical state, be it liquid or finely granulated. Methods for accomplishing these objectives have been fairly well worked out and have been used for many years. Probably improvements can be made. The requirements for stability of honey are more stringent now than in the past, with honey a world commodity and available in supermarkets the year around. Government price support and loan operations require storage of honey, and market conditions also may require storage at any point in the handling chain, including the producer, packer, wholesaler, and exporter.

The primary operation in the processing of honey is the application and control of heat. If we consider storage to be the application of or exposure to low amounts of heat over long periods, it can be seen that a study of the effects of heat on honey quality can have a wide application.

Any assessment of honey quality must include flavor considerations. The objective measurement of changes in flavor, particularly where they are gradual, is most difficult. We have measured the accumulation of a decomposition product of the sugars (hydroxymethylfurfural or HMF) as an index of heat-induced chemical change in the honey. Changes in flavor, other than simple loss by evaporation, also may be considered heat-induced chemical changes.

To study the effects of treatment on honey, we must use some properties of honey as indices of change. Such properties should relate to the quality or commercial value of honey. The occurrence of granulation of liquid honey, liquefaction or softening of granulated honey, and

fermentation as functions of storage conditions has been reported; also, color is easily measured.

As indicators of the acceptability of honey for table use, Europeans have for many years used the amount of certain enzymes and HMF in honey. They considered that heating honey sufficiently to destroy or greatly lower its enzyme content or produce HMF reduced its desirability for most uses. A considerable difference has been noted in the reports by various workers on the sensitivity to heat of enzymes, largely diastase and invertase, in honey. Only recently has it been noted that storage alone is sufficient to reduce enzyme content and produce HMF in honey. Since some honey types frequently exported to Europe are naturally low in diastase, the response of diastase and invertase to storage and processing is of great importance for exporters.

A study was made of the effects of heating and storage on honey quality and was based on the results with three types of honey stored at six temperatures for 2 years. The results were used to obtain predictions of the quality life of honey under any storage conditions. The following information is typical of the calculations based on this work.

At 68°F, diastase in honey has a half-life of 1,500 days, nearly 4 years. Invertase is more heat sensitive, with a half-life at 68° of 800 days, or about 2¼ years. Thus there are no problems here. By increasing the storage temperature to 77°, half the diastase is gone in 540 days, or 1⅛ years, and half the invertase disappears in 250 days, or about 8 months. These periods are still rather long and there would seem to be nothing to be concerned about. However, temperatures in the 90's for extended periods are not at all uncommon: 126 days (4 months) will destroy half the diastase and about 50 days (2 months) will eliminate half the invertase. As the temperature increases, the periods involved become shorter and shorter until the processing temperatures are reached. At 130°, 2½ days would account for half the diastase and in 13 hours half the invertase is gone.

A recommended temperature for pasteurization of honey is 145°F for 30 minutes. At this temperature diastase has a half-life of 16 hours and invertase only 3 hours. At first glance this might seem to present no problems, but it must be remembered that unless flash heating and immediate cooling are used, many hours will be required for a batch of honey to cool from 145° to a safe temperature.

If we proceed further to a temperature often recommended for preventing granulation, 160°F for 30 minutes, the necessity of prompt cooling becomes highly important. At 160°, 2½ hours will destroy half of the diastase, but half of the more sensitive invertase will be lost in 40 minutes. This treatment then cannot be recommended for any honey in which a good enzyme level is needed, as for export.

The damage done to honey by heating and by storage is the same. For the lower storage temperatures, simply a much longer time is required to obtain the same result. It must be remembered that the effects of processing and storage are additive. It is for this reason that proper storage is so important. A few periods of hot weather can offset the benefits of months of cool storage—10 days at 90°F are equivalent to 100 to 120 days at 70°. An hour at 145° in processing will cause changes equivalent to 40 days' storage at 77°.

An easy way for beekeepers to decide whether they have storage or processing deterioration is to take samples of the fresh honey, being careful that the samples are fairly representative of the batch, and place them in a freezer for the entire period. At the end of this time, they should warm the samples to room temperature and compare them by color, flavor, and aroma with the honey in common storage. In some parts of the United States, the value of the difference can reach 1½ cents per pound in a few months. Such figures certainly would justify expenditures for temperature control.

People who store honey are in a dilemma. They must select conditions that will minimize fermentation, undesirable granulation, and heat damage. Fermentation is strongly retarded below 50°F and above 100°. Granulation is accelerated between 55° and 60° and initiated by fluctuation at 50° to 55°. The best condition for storing unpasteurized honey seems to be below 50°, or winter temperatures over much of the United States. Warming above this range in the spring can initiate active fermentation in such honey, which is usually granulated and thus even more susceptible.

References

Doner, L. W.
  1977. THE SUGARS OF HONEY—A REVIEW. Journal of Science and Agriculture.
Townsend, G. F.
  1961. PREPARATION OF HONEY FOR MARKET. 24 p. Ontario Department of Agriculture Publication 544.
White, J. W. Jr.
  1975. HONEY. In Grout, R. A., ed., The hive and the honey bee, p. 491–530. Dadant & Sons, Inc., Hamilton, Ill.

——
  1975. COMPOSITION AND PHYSICAL PROPERTIES OF HONEY. In E. Crane, ed., Honey Review, p. 157–239. Heinemann, London.
——— M. L. Riethop, M. H. Subers, and I. Kushnir.
  1962. COMPOSITION OF AMERICAN HONEYS. 124 p. U.S. Department of Agriculture Technical Bulletin 1261.

HONEY REMOVAL, PROCESSING, AND PACKING

By B. F. Detroy [1]

Honey is at its peak quality when properly cured and sealed in the comb by the honey bee. When it is converted from this state by humans to suit their particular needs, deterioration begins. The extent of deterioration depends on the processing methods used between the time the honey is extracted from the comb and its use by the consumer. It is the responsibility of the industry to provide a top quality product to the consumer if acceptance is to be expected.

Since most honey harvested is extracted from the combs in an extracting plant, the beekeeper should equip this plant so that the operations can be done in the most efficient manner possible to provide a high-quality product for market.

Honey House

The extracting plant is generally located in a honey house. The honey house is the center of activities for beekeepers, represents a goodly portion of their investment, and may contribute greatly to the overall efficiency of their entire operation.

The honey house may contain various other facilities in addition to the extracting plant, such as storage space for hive equipment and honey, workshops, office space, and possibly a packing or salesroom or both. The building should be designed for the work to be done in it and be properly equipped. Efficient arrangement, cleanliness, and ample space are of prime consideration in planning the honey house.

Types of Honey Houses

Honey houses may be one- or two-story structures. The one-story structure probably is the most common and is used by both large and small beekeepers. The building construction is simple, and the choice of a building site is not limited by terrain. Honey handling cannot be readily adapted

to gravity flow, but proper use of honey pumps can overcome this disadvantage. Small beekeepers may use this type of building to particular advantage. Since all equipment can be compactly arranged on one level, it is easier to closely regulate all operations.

Two-story structures may have both floors above ground or may have one floor above ground and a full basement. If both floors are above ground, it is necessary to provide a ramp to the second floor so that it is accessible to trucks or to install elevators to move equipment and material from one floor to the other. A hillside building site will provide access to both floors without ramps and both floors are at least partially above ground (fig. 1). The main advantage of the two-story building is that the supers of honey can be unloaded at the upper level where the extracting plant is located. Extracted honey can then flow by gravity to storage or further processing on the lower flow.

Space Requirements

Careful planning before building a honey house may save costly additions later. The operations necessary in extracting honey and the sequence in which they are performed should be considered in detail for filling the needs of the beekeeper. Ample space should be provided for all extracting and processing equipment. The equipment chosen should be expected to operate at near-rated capacity and be compactly arranged so that the material flows smoothly from operation to operation with a minimum of movement by the operators from area to area.

Storage space for full honey supers should be figured on the basis of the maximum number anticipated in the honey house at any one time. Uniform stack height of full supers should be used throughout the operation for efficient handling. Warm storage areas for full honey supers aid in the extracting operation. An area for storage of liquid extracted honey also must be provided. Required space should be figured on the basis of

[1] Agricultural engineer, Science and Education Administration, Bee Management and Entomology Research, University of Wisconsin, Madison, Wis. 53706.

FIGURE 1.—Apiculture laboratory and honey house at University of Wisconsin. Note that lower floor is partly below ground and entrances are at both floor levels.

the type of bulk containers used and the height they are to be stacked.

Other areas to be included in the building will vary in type and size according to individual beekeepers and the operations they desire to perform. They may include a room for rendering cappings, space for shop facilities, equipment-assembling area, truck storage, small packing plant, salesroom, and office space.

Special Features

Regardless of type or size, the building should be bee-tight and provide a means of unloading filled honey supers in a closed-in area. A sunken driveway or raised platform to permit loading or unloading with the truck bed at floor level will prove to be a great labor and time saver.

The floors, walls, and ceiling should be made of materials that can be easily cleaned and maintained. Walls and ceilings should be a light color and of a material or paint that can be washed often. Floors should be of concrete where possible. If hardwood floors are necessary, they should be covered with ceramic tile. Any floor used must withstand heavy loads and be free of vibration. A smooth surface with ample drain facilities will make cleaning easier.

Illumination and ventilation should be carefully considered in planning the building. In areas where close operator attention is required, it is highly important to provide adequate illumination, 50 to 75 foot-candles. The entire building should have adequate lighting. Windows to provide both light and ventilation should be strategically located. Fans may be installed to reduce odors and lower humidity. All windows must be screened and provided with bee escapes.

Rooms in which fires might originate, such as the furnace room or wax-rendering room, should be built of fire-resistant material or lined with asbestos board. It may be desirable to have such facilities in another building, and if so, the buildings should be separated by at least 20 feet.

If handtrucks, motorized lifts, and fork or barrel trucks are to be used, doorways should be large enough to permit their free movement to all parts of the building.

Removing Honey Supers

Honey supers should be removed from the hive as soon as the honey is sealed. Extraction soon after removal may prevent crystallization in the comb. It often is possible to reuse the super on the colony before the flow ceases.

For the small beekeeper, the simplest way to remove bees from the honey super is to brush them off each comb. Another method of removing bees from supers is to use an inner cover with a bee escape. The escape board, as it is called, is placed between the supers and the brood nest the day before the supers are to be removed. All other openings into the supers must be closed so that the bees cannot return. This method works best in cool weather when the bees move at night from the honeycombs to the brood nest.

A recent innovation for removing bees from supers of honey is to blow them out. A large vacuum cleaner with a crevice tool attachment works satisfactorily. The larger household tank-type vacuum cleaner can be used if the dust bag is removed. A velocity of about 18,000 feet per minute is required. There are now engine-driven blowers built especially for removing bees. This method does not irritate the bees and it is effective.

Care and Storage of Supers

Supers are most easily handled if they are kept in uniform stack heights from the time the full supers are removed from the colony until the empties are returned. Stack boards or skid boards are commonly used (fig. 2). The boards are frequently fitted with a sheet metal pan or tray to catch any honey drip that may come from the stack of supers.

Skid boards placed on the truck bed receive the supers as they are removed from the colonies and moved through the entire extracting process and back into the truck with an ordinary warehouse truck (fig. 2).

More expensive types of lift trucks are sometimes used that can handle skids containing two or four stacks of supers. Castered dollies also may be used, but these are not easily hauled to and from the yards on trucks.

Hot rooms are sometimes used for storage of filled supers before extraction, expecially in regions where cool temperature or high humidity is common. If the only requirement is to keep the honey warm to facilitate extraction, the room should be kept at 75° to 100°F and have a circulating fan. Warm, dry air also may be used to remove moisture from the honey, in which case slatted skid boards should be used or the supers stacked crisscross. The air should be introduced at floor level and drawn up through the stacks. Moisture-laden air should be discharged from the room.

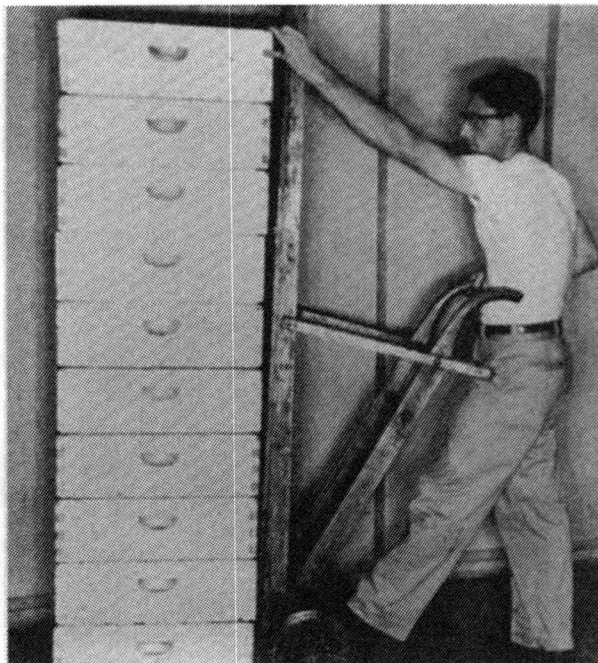

FIGURE 2.—Equipment for moving supers: *Top*, One-man hand truck with skid board can move several at a time; *bottom*, power equipment handles many more supers in stacks, as well as drums of honey.

Dehumidifiers also may be used to speed the lowering of the moisture in the honey.

Supers stored for a long time should be fumigated to prevent damage by the larvae of the wax moth.

Honey Extraction

Equipment in the extracting area should be arranged so that the material flows smoothly through the various operations with a minimum of interruption and with as little physical effort as possible.

Extracting equipment will differ in almost every honey house. Choice of equipment is dependent on the size of operation, physical properties of the honey, availability of labor supply, and the personal selection of the individual. Many beekeepers have designed and built their own equipment or remodeled commercial equipment to meet some particular need.

Uncapping Devices

To extract honey, it is first necessary to remove the capping from the comb cells. A wide range of equipment is available for uncapping combs, from unheated hand knives to elaborate mechanical machines.

A cold knife can be used to uncap warm combs or it can be heated by placing it in hot water. It is most commonly used by the hobbyist beekeeper who has only a few hives. The steam and electrically heated hand knife and hand plane are probably the most widely used uncapping devices in this country today. In a more refined version, the knife is mounted in a frame on spring steel mounts and vibrated by an electric motor. This type, referred to as the jiggler knife, may be mounted in a vertical, horizontal, or inclined position. The knife vibrates in the direction of its length.

Machines that carry the combs of honey through an uncapping device and sometimes into the extractor after being fed into the machine by hand are available commercially and their use is becoming more common (fig. 3). The uncapping devices on these machines may be vibrating knives, rotating knives, flails of various kinds, or perforating rollers. A few machines have been built by individuals that uncap the frames in the super.

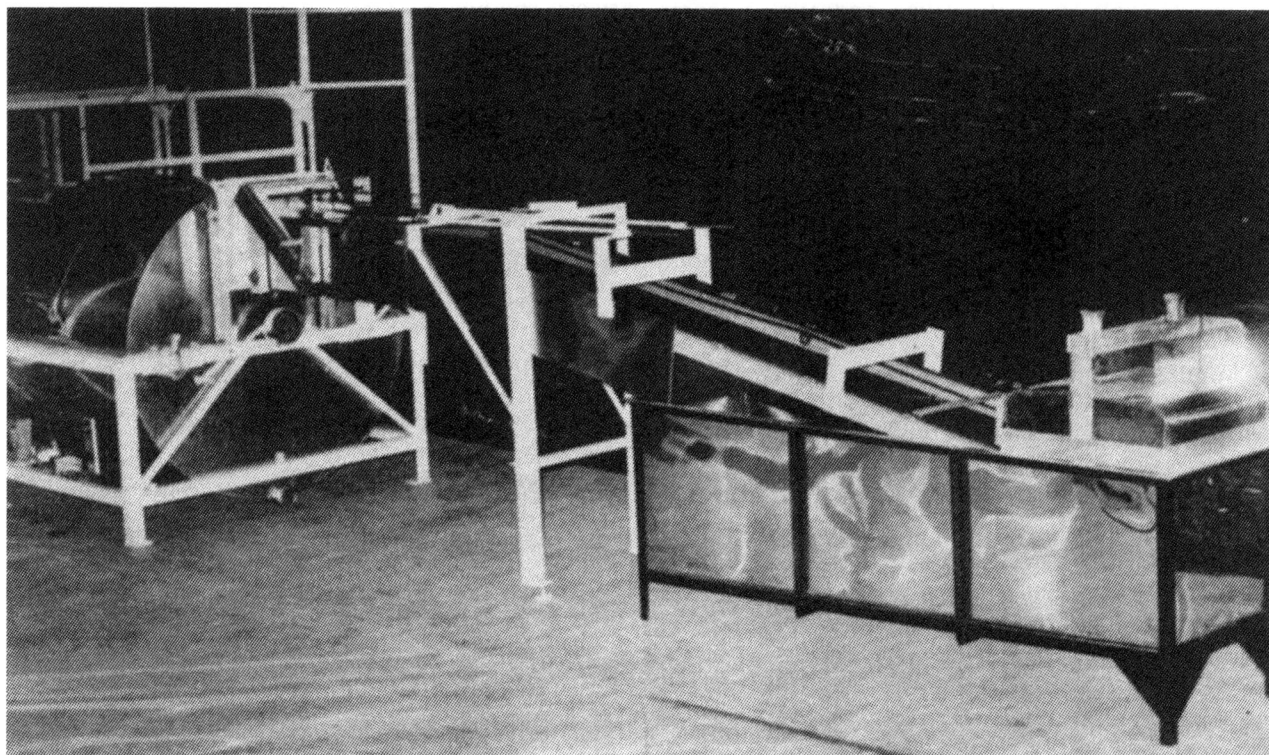

FIGURE 3.—Mechanical uncapping machine with uncapped frame conveyor and automatic load extractor.

Extractors

Two types of extractors are in use in this country today—the reversible basket and the radial. Both use centrifugal force to remove the honey from the comb.

Reversible-basket extractors range in size from 2 to 16 frames per load. Honey is extracted by applying centrifugal force to first one side of the comb then the other. The comb is reversed three or four times—turned 180°— during the extracting cycle. On some of the small extractors, the frames are reversed by hand with the machine stopped, whereas on others the frames are reversed by use of a brake while the machine is running. Extracting time ranges from 2 to 4 minutes at constant speed.

Radial extractors range in size from 12 to 80 frames per load. Both sides of the comb are extracted simultaneously as the combs are rotated, the centrifugal force acting radially across the face of the comb. The extracting cycle is started at 150 revolutions per minute and is gradually increased during the cycle to 300 revolutions per minute. The time required to extract a load of combs will vary from 12 to 20 minutes, depending on the temperature and density of the honey.

Special extractors have been built and used that are larger than those described and that extract the combs in the supers, special boxes, or baskets.

Automatic electric controls have been developed for both the radial and reversible-basket extractors. These controls change the revolutions per minute, reverse the baskets, and shut off the motor when the cycle is completed. Mechanical controls also are available that automatically increase the speed of the radial extractor during the extracting cycle.

Care of Cappings

Cappings and honey removed from the combs in the uncapping operation must be separated to salvage the honey and wax. Caution must be taken in recovering the honey to prevent impairing the flavor, color, and aroma. The following methods are used:

(1) *Draining by gravity.*—The cappings are accumulated in screened or perforated containers and allowed to drain, usually for at least 24 hours in a warm room. Stirring and breaking up the cappings facilitate draining.

(2) *Centrifuging.*—The cappings are placed in a specially constructed centrifugal drier or in wire baskets that fit into a radial extractor. Honey is removed from the cappings by centrifugal force as the cappings rotate.

(3) *Pressing.*—A basket-type perforated container is used to catch the cappings where some gravity draining takes place prior to pressing. Usually the container is placed directly under the press ram and pressure applied to squeeze the honey from the cappings.

Honey removed from the cappings by any of these methods is undamaged. Usually the remaining cappings will contain as much as 50 percent honey by weight, which may be recovered when the cappings are melted. This honey is generally damaged by overheating and should be handled separately.

(4) *Flotation and melting.*—The cappings' melter is widely used to separate honey and cappings. The cappings and honey enter the melter tank near the bottom and are separated by gravity. Separation is facilitated by heat that softens the cappings and increases the fluidity of the honey. The cappings being less dense rise to the top where they are melted. The honey level is controlled by an adjustable height overflow enclosed by a baffle to prevent the entry of wax. A layer of cappings in various stages of liquefaction is maintained between the honey level and the heat source. Liquid wax accumulates at the top of the tank and is discharged into solidifying containers. Heat may be supplied by steam coils, electric heaters, heat lamps, or radiant gas heaters.

Various models of this type of separator are marketed. If they are properly operated, the honey obtained can be added to the remainder of the crop without damage to grade, color, or flavor.

(5) *Centrifugal separator.*—Recent development of a centrifugal separator (fig. 4) that automatically separates the honey and dries the cappings has greatly advanced the use of mechanical uncappers. Usually all honey and cappings from the uncapper and extractor are run through the machine. Large pieces of cappings should be broken up to assure proper feeding into the separator.

Processing

Processing the honey crop beyond the extraction stage may be done by the producer, the packer, or both. Regardless of where these operations take place, they are necessary to provide the consumer with a high-quality product. It is important, however, that the heating be con-

FIGURE 4.—Centrifugal separator for automatically separating cappings from honey and drying the cappings.

trolled, since the flavor, color, and aroma of honey can be seriously impaired by excessive temperature over a given period of time.

The Sump and Pump

Honey from the uncapping and extracting operation usually flows into a sump. The sump is a tank, usually water jacketed, that collects honey from the extracting process so that it can be delivered for further processing at a uniform rate. The sump may contain a series of baffles or screens or both for removing coarse wax particles and other foreign material. A honey pump is generally used in conjunction with the sump; however, in some systems, gravity flow can be used and the pump eliminated.

Gear pumps or vane pumps are commonly used. Where the centrifugal separator is used and it is necessary to pump large quantities of cappings, some other type of pump may be required. Pumps used in a continuous flow system should be supplied with honey in sufficient quantity to allow uninterrupted operation. To prevent introduction of air into the honey, the pump should run at low speed and the level of honey in the sump kept well above the pump intake. Automatic pump con-

trols, either float type or electric liquid level control type, can be used to eliminate continual operator supervision.

Strainers

After the bulk of the wax has been removed from the honey by the sump tank, coarse straining, or centrifugal separators, it is necessary to remove very fine material. Settling of honey may prove satisfactory for some processors. The honey is first screened in a sump and then pumped into settling tanks at a temperature of at least 100°F. Sufficient time should be allowed to permit the required separation.

To be certain that all honey packed will meet the desired grade requirements, it is necessary to use some type of strainer. Many types and sizes are used and the straining media may be metal screen, crushed granite, silica sand, or cloth. Regardless of the material used, the mesh must be fine enough to produce the desired result. Cloth has the advantage of being easily cleaned; furthermore, since the initial cost is low, a cloth may be used only once and discarded.

Honey may be moved through the strainer by pressure (pumping) or by gravity flow. When cloth strainers are used in a pressure system (fig. 5), a pressure switch should be installed in the honey line to prevent excessive pressure that could rupture the strainer cloth.

Heating the honey to 115° will greatly facilitate the straining process. This increases the fluidity of the honey without softening the wax particles appreciably. Higher temperatures will soften the wax so that it may be forced into or through the straining media.

Heating and Cooling

Heat properly applied can be a great aid in handling honey. Heat also dissolves coarse crystals and destroys yeasts, and thus prevents fermentation and retards granulation. Heat may also seriously damage the color, flavor, and aroma of honey unless particular precautions are taken. Damage may result from a small amount of heat over long periods of time as well as high temperatures for short periods of time.

Several methods of heating are used successfully. Shallow pans with inclined surfaces heated by water jackets are commonly used. As the honey flows into the pan, it should be distributed over the surface by suitable baffles. Jacketed tanks may be used for heating, in which case the honey should

FIGURE 5.—Cloth strainer showing containing shell, supporting framework with strainer cloth, gasket, and coverplate.

be slowly but continuously agitated to ensure uniform heating throughout the tank. Heat exchangers in which the honey is pumped quickly through a passage contained in hot water are used very successfully as flash heaters.

One design of heat exchanger is shown in figure 6. This exchanger consists of three concentric tubes in which the honey is pumped through a ³⁄₁₆-inch-thick annular space between two layers of flowing hot water. Honey enters the exchanger at the bottom and is pumped in a direction opposite to the flow of water. These units may be connected in series to provide the desired amount of heating.

Precautions for cooling honey after heating are seldom practiced to a suitable degree. Immediate cooling following flash heating is essential to prevent honey damage. Equipment similar to that used for heating can be used effectively for cooling by using cold water instead of hot. Heat ex-

changers are particularly effective, but may cause excessive line pressure as the honey becomes more viscous upon cooling.

Storage of Honey

Honey in bulk containers, 60-pound cans, or 55-gallon drums should be stored in a dry place at as near 70°F as possible. Long periods of storage above 70° will damage the honey the same as excessive heating. Storage of unheated honey at 50° to 70° is inducive to granulation and fermentation. This also is true for honey packed in bottles and other small containers. These should be stored in shipping cases to protect them from light.

Most deterioration of honey during storage can be prevented by maintaining storage temperatures below 50°. Honey stored at freezer temperatures, 0° to −10°, for years cannot be distinguished from fresh extracted honey in color, flavor, or aroma.

FIGURE 6.—Concentric tube heat exchangers connected in series and used for flash-heating honey. (Note strainers ahead of flash heaters.)

Handling Honey

Honey is usually sold wholesale to packers in 5-gallon cans, drums, or in bulk. For many years, the 5-gallon (60-pound) can was the principal wholesale container, and small beekeepers still use it. Although single cans may be handled by hand, they are more easily handled by pallet truck. Larger quantities of honey are more easily handled in 55-gallon drums. The drums are more durable than cans and are reusable. Several industrial hand and power trucks are available for handling drums. If drums are to be stacked, a motor-driven lift truck is needed. All commercial bottlers of honey are equipped to handle both cans and drums.

Some beekeepers have built large tanks for storage of 1,000 gallons or more of honey, which is then pumped into tanks on the trucks for transfer to bottlers. Some companies use air pressure to transfer honey into and out of the truck tanks.

Producer Marketing

Producers have a choice of methods for disposing of their honey crop. They may sell their entire crop in bulk containers to a packer or dealer or they may pack a part or all of it and sell direct to retail stores or consumers or both. Producers may be members of a cooperative through which their honey is processed and sold.

Wholesale Marketing

Producers who market honey in bulk should keep in mind the market to be supplied when choosing the type of container to use. Generally, these containers will be either the 60-pound can or the 55-gallon drum. A limited quantity of honey is moved from the producer to the packing plant in tank trailers. Careful sampling is necessary when the honey is extracted. Representative samples should be taken from each tank, each yard, or each day's run and should be carefully marked on both the sample and the containers for accurate identification.

Honey generally is sold at wholesale prices on the basis of samples, and accurate sampling will result in building confidence, understanding, and satisfaction for both producer and buyer. Producers who know exactly what they have for sale can demand and get top market prices. Packers who know exactly what they buy can readily process and blend to meet their particular

standards without concern for discrepancies or variation.

Approximately 50 percent of the honey produced in the United States is marketed by the producer in bulk.

Producer-Packers

Honey producers who bottle and sell part or all of their honey crop are referred to as producer-packers. Almost half of the honey produced in the United States is marketed in this manner.

Producer-packers receive a higher price per pound for their honey; however, they may have many additional costs. Processing equipment that will yield a product meeting the desired grade standards must be used. The honey must compete with other brands of honey and other foods backed by aggressive sales and promotion programs. They have to employ a broker to move the honey into retail channels.

Many producer-packers confine their sales to salesrooms in their homes or honey houses, roadside stands, door-to-door sales, or local stores. Some have established regular sales routes to supply retailers over a wide area, and these routes are serviced at regular intervals.

Cooperative Marketing

There are several cooperative marketing organizations in the United States. These organizations may buy the member producer's crop and process, pack, and distribute the products under the cooperative label. Other organizations may only pool and market the member's production in bulk containers.

Generally, the cooperative may operate as follows: Member producers are furnished with bulk containers. When the crop is harvested, the honey is put into these containers and shipped or trucked to the cooperative by the producers. The honey is then graded and the producers are paid a part of the total price. The cooperative then processes, packs, and sells the honey through its sales organization. At a later date, producers are paid the remainder of their selling price.

Cooperative marketing offers many advantages, but there also are some disadvantages, just as in other types of marketing. Producers must decide which method of marketing is the most advantageous to them and market their crop accordingly.

Packing

Honey packed for market must be of high quality, neatly packaged in clean, attractive containers, and attractively labeled. Every caution should be taken in processing and packing to ensure a product of quality as near as possible to that sealed in the cell by the bee. All honey packed under a given label should be as uniform as possible to assure consumer satisfaction. An attractive, eye-catching display in a prominent location is desirable.

Most large honey packers have automatic labeling, filling, and capping equipment. Their honey is distributed and sold under their advertised brand, usually in a limited area. Few, if any, have nationwide distribution. Some have sales personnel, whereas others employ food brokers or other sales agencies to market their product. Many use warehousing facilities in areas of concentrated retail outlets.

Much honey is sold in bulk for industrial consumption, such as for the baking industry, restaurant trade, honey candies, and honey butter. Many other industries use honey in varying quantities.

Liquid Honey

Liquid honey is packed in glass, tin, plastic, and paper containers. Glass is the most popular and is used in a wide variety of shapes and sizes. Plastic containers in various shapes are becoming more and more popular. The 12-ounce plastic container makes a very satisfactory table dispenser. Special glass and plastic containers are used effectively in novelty and gift packs and are popular on the retail market.

Bottled honey should be free of air bubbles or any foreign particles and the containers must be spotlessly clean. Honey bottled by floral source should be clearly labeled as such to ensure customer satisfaction.

Honey selected for bottling should be from floral sources that granulate slowly. Proper heating in the processing and bottling operation also will help retard granulation. Commercial packing plants put much of the honey prepared for the liquid honey trade through a pressure-filter process. Any bottled honey in a sales display that shows signs of granulation should be replaced immediately.

Granulated or Creamed Honey

The popularity of granulated or creamed honey is increasing in the United States. This honey is presently available in many retail food stores. It is packed in various paper, plastic, and glass containers. The desired consistency of creamed honey is soft and smooth to allow easy spreading at room temperature.

Honey used for this purpose should be from a floral source that granulates rapidly into a product of soft, smooth, fine, creamy consistency. Honeys that granulate slowly may be used by adding about 10 percent of finely ground crystallized honey. To encourage granulation, the honey should be refrigerated immediately after the fine honey crystals have been added to prevent any air bubbles rising to the surface. After rapid cooling, the honey should be stored at 55° to 57°F and that temperature maintained until the honey is completely crystallized. Cool storage is desirable.

Creamed honey will remain firm at room temperature, but will break down if subjected to high temperature or high humidity. Once it has softened or partially liquefied, recooling will not make it firm again.

Comb Honey

Comb honey is marketed in the form of section comb, cut-comb, and chunk. All forms require special care and handling, and when properly prepared they have excellent consumer appeal.

Section comb honey is produced in a special super. When removed, the sections are carefully scraped with a suitable instrument to remove the propolis. The sections are then sorted, graded, and placed in window cartons or wrapped in cellophane. Some packers seal the sections in clear plastic bags before placing them in window cartons. Sections that do not meet the required grade standards should not be marketed.

Cut-comb honey is produced in shallow supers on foundation similar to that used for sections but in frames instead of sections. The comb honey is cut from the frames into the desired size for marketing. Sizes of cut-comb honey vary from a 2-ounce individual serving to large pieces weighing nearly a pound. The cut edges of the comb must be drained or dried in a special centrifugal drier, so that no liquid honey remains. The pieces are either wrapped in cellophane or heat sealed in polyethylene bags and packaged in containers of various styles.

One of the most attractive and appealing packs of honey is the chunk honey pack. It consists of a chunk of comb honey in a glass container surrounded with liquid honey.

When packing chunk honey, the pieces of well-drained cut-comb are placed in the container, usually glass jars. The containers then are filled with liquid honey that has been heated to retard granulation. The liquid honey should be run down the inside of the container to prevent introduction of air bubbles and should be at a temperature of 120°F. The containers should be capped and laid on their sides immediately after filling to prevent damage to the comb because of its buoyancy.

Special widemouthed jars are used for packing chunk honey. The chunk should be as wide as possible and still slip readily into the jar, and the length should extend from the top to the bottom of the container.

Probably the greatest deterrent to packing chunk honey is the tendency of the liquid honey to granulate.

References

GROUT, R. A., ed.
    1975. THE HIVE AND THE HONEY BEE. Revised. 739 p. Dadant & Sons, Hamilton, Ill.
OWENS, C. D. and B. F. DETROY.
    1965. SELECTING AND OPERATING BEEKEEPING EQUIPMENT. 24 p. U.S. Department of Agriculture, Farmers' Bulletin 220.
TOWNSEND, G. F.
    1961. PREPARATION OF HONEY FOR MARKET. 24 p. Ontario Department of Agriculture Publication 544.

SHOWING HONEY AT FAIRS

By E. C. Martin [1]

Showing farm produce at the county fair or the State fair is a fine American tradition (fig. 1). Fair visitors can be so fascinated by attractive displays of honey and other apiary products, including observation hives, that we should surely make greater use of such opportunities to promote our products. Honey consumption in the United States is only slightly more than 1 pound per person. If no effort is made to promote its use, consumption could drop still further—and there could be a tendency toward lower prices.

About 200,000 people keep bees in the United States. Most States have a fair and there are hundreds of county fairs. Beekeepers in some States do marvelous jobs of organizing displays at the fairs. The initiative for getting beekeepers' displays on the fair prize list and then stimulating good, competitive response from the honey producers must come from State or local beekeepers' associations. Persistent effort by a continuing committee can develop the talent for showmanship present in every community. Expanded use of the fairs could provide the beekeeping industry with an interesting and profitable way to tell many millions of people the good qualities of honey.

Honey Festivals

Ohio beekeepers have for some years worked with the Chamber of Commerce of a strategically located town to stage an annual honey festival lasting several days. Beekeeping displays, educational features, and booths where honey is sold are located in the streets and buildings in the center of town. Local sports events, band concerts, and other activities are all part of the festival. Thousands of people enjoy the festival and much honey is sold. More recently, Michigan beekeepers have followed Ohio's lead and other States have shown interest.

[1] Staff scientist, National Program Staff, Science and Education Administration, Beltsville Agricultural Research Center-West, Beltsville, Md. 20705.

FIGURE 1.—Large observation hive on display at a State fair. Access to the outdoors was provided by a clear plastic pipe.

The Prize List

In working with county or State fairs, beekeepers have to make practical decisions on the most appropriate competitive classes that will best display locally produced products and attract the largest number of competitors (fig. 2). Some State fairs may require an entry of 100 pounds or more

103

FIGURE 2.—Part of a display at a State fair. Cabin is made of beeswax.

in classes such as white extracted honey. This might be too much for good competition at a small fair where an entry of six 1-pound jars might be sufficient.

Following are classes from which selections can be made, as appropriate to the area and the type of fair:

1. Display of apiary products. Specify the floor or table space available for each exhibitor and the required products.
2. Liquid honey with separate classes for major color types. Specify the number and size of jars for each entry.
3. Finely granulated (creamed) honey.
4. Section comb honey.
5. Cut comb honey.
6. Chunk honey (a piece of comb honey in a jar of liquid). Specify the number and size of jars to be entered.
7. Beeswax blocks of specified weight and possibly an artistic display of molded or carved beeswax figures.
8. Langstroth extracting combs of honey of specified color.

9. One-frame observation hive containing worker bees, drones, a queen bee, and brood.
10. A collection of pressed specimens of honey and pollen plants.
11. Special classes of honey-baked products such as cakes, cookies, and preserves.

Objectives and Suggestions

1. Besides promoting the sale of honey, showing honey at fairs gives many beekeepers experience in preparing an ideal pack of honey in its most attractive form.

2. If at all possible, persuade the fair board to allow beekeepers to sell honey from their competitive booths. This can mean the difference between success and failure in the whole effort.

3. Try to avoid occasions for hard feelings among competitors. Judging can be made less personal if entries are unlabeled and brought to the judges so that they do not know the owners of the products.

4. Make the requirements for the classes in the prize list specific and complete. Aim for uniformity

within classes. If requirements are haphazard and such things as size of container and color of honey are not clearly specified, the diverse entries cannot be properly judged and entries will not look attractive to the visiting public.

5. Don't allow filtered honey in a nonfiltered honey class.

6. Have a class for smoothly granulated honey. Consumers need more information on the fact that this is another good form in which to eat honey.

7. Use advertising statements in displays that are bright, enthusiastic, and educational without straying into unproved statements about medical or nutritive properties of honey.

8. Encourage local newspapers to interview and report the winners and publish pictures of the displays.

9. Encourage special displays by 4–H Club or other young people's groups.

10. Try to have the national honey queen in attendance.

Preparing Honey for Competition

Competition—particularly in white, liquid honey classes—can become quite keen, and some beekeepers become very expert in preparing honey for shows. Where competition is keen, beekeepers sometimes select the most ideal combs of honey, extract them in a hand extractor without the use of a honey pump so as to avoid incorporating air bubbles, strain the honey carefully and allow it to settle, and place it in jars free from crystals, bubbles, or specks of any kind. If show honey contains crystals, the honey may be heated cautiously until the crystals dissolve. Air bubbles may be brought to the surface by gently warming the honey for an hour or more. Moisture is best removed from honey by exposing combs to warm, dry, moving air before extracting.

Smoothly granulated honey is prepared for shows by seeding liquid honey with about 10 percent finely crystallized honey, mixing carefully, bottling, and storing at a temperature as close to 57°F as possible. If stored at the right temperature, the prepared honey will set firmly in about a week.

Judging Honey

It is sometimes difficult to secure the services of a competent judge, so it is wise to publish a fairly comprehensive score card in the prize list to help the judge consider all appropriate points. Following are some suggestions for judges:

1. Take along a container for water, several cloths, a drinking glass, toothpicks for tasting samples, an extension cord and light bulb to backlight samples, pencils, and record sheets.

2. At all major fairs, use a refractometer to measure honey water content and a Pfund grader or color classifier to measure the color of liquid honey entries.

3. Have samples in each class brought to a judging table. Keep arranging and rearranging the samples in order of merit and after some study the relative merits will stand out clearly.

4. Follow the rules and score card, but don't be overly strict at smaller fairs where beginners need encouragement.

5. Beekeepers put their greatest effort into preparing displays, and prize money is greatest in this category so the competition may be keen. Points often are divided between the quality and appearance of the apiary products in the display and the educational or advertising value and originality exhibited. Some form of motion such as the use of a turntable containing a pyramid of honey attracts attention and earns points.

The following score card is more detailed than most, but because of this it should be more helpful in reminding judges of the important points to keep in mind when judging, and competitors of the more important points to keep in mind when preparing their entries. If judging standards are provided in the fair prize list, the judge should follow them.

Liquid Honey

	Points
Appearance, suitability, and uniformity of containers	5
Uniform and accurate volume of honey	5
Freedom from crystals	10
Freedom from impurities, including froth	20
Uniform honey in all containers of the entry	5
Color	10
Brightness	10
Flavor and aroma	15
Density (No additional points below 16 percent water)	20
	100

Granulated (Creamed) Honey   *Points*

Appearance, suitability, and uniformity of containers	5
Uniform and accurate volume of honey	5
Firmness of set (not runny but spreadable)	20
Texture of granulation (smooth and fine)	20
Absence of impurities, including froth	15
Uniform honey in all containers of the entry	10
Color	10
Flavor and aroma (such as natural flavors present and undamaged by heat)	15
	100

Comb Honey in Standard Sections

Suitability, uniformity, and cleanliness of sections (wood)	20
Completeness, uniformity, and cleanliness of cappings	30
Uniform and completely filled honey cells	30
Quantity, quality, and uniformity of honey	20
	100

Cut Comb Honey

Accuracy and neatness of the cut edge of the comb	20
Uniform depth and filling of the honey cells	20
Complete, uniform, and clean cappings	20
Quality, quantity, and uniformity of honey	20
Freedom from leakage and general appearance of the pack	20
	100

Chunk Honey

Uniformity, cleanliness, and general appearance of the entry	30
Freedom from impurities and granulation	20
Quality of the liquid honey	25
Quality and neatness of the comb honey	20
Uniform and accurate volume of honey	5
	100

Beeswax   *Points*

Color between straw and canary yellow (such as undamaged by propolis and iron stain)	30
Cleanliness (free from surface dirt, honey, and impurities)	25
Uniform appearance of all wax in the entry	15
Freedom from cracking, shrinkage, and marks	15
Texture and aroma (such as pure wax free from hard water damage)	15
	100

Bees in an Observation Hive

Correct type and color of bees for the class	15
Queen: Size, shape, and behavior	15
Brood pattern	15
Variety: Presence of queen, workers, drones, brood, honey, pollen, and so forth	15
Correct number of bees for interest and ease of observation	10
Cleanliness and suitability of the combs	15
Appearance, cleanliness, and suitability of the observation hive	15
	100

Display of Apiary Products

Educational value	20
Advertising value (normally for the products in general, not a brand)	20
Attractive arrangement (pleasing and eye-catching)	20
Originality and variety	20
Appearance and quality of products in the display	20
	100

POLLINATION OF CROPS

By S. E. McGregor[1]

Farmers can provide the best agronomic practices—proper seedbed preparation, fertilization, soil moisture, cultivation, pest control, and harvesting methods—and fail to obtain a bountiful harvest, if they neglect to provide for pollination. Many of their fruit crops, legumes, vegetables, and oilseed crops depend upon pollinating insects for set of fruit or seeds. Therein lies the basis for the most important contribution made by any insect to agriculture—pollination by the honey bee.

Flowering and Fruiting in Plants

Some knowledge of the flower structure and fruit setting is necessary to understand the entire pollination process. All flowers have the same basic pattern, but there are many variations. The peach blossom, the tassel and ear of corn, and the sunflower head appear remarkably different, but all have the same basic parts.

Typically, the flower (fig. 1) is composed of the sexual organs, protected by the usually colorful delicate *petals* that may form a sort of tube or crownlike *corolla*. These are supported and partially protected by the usually green and more durable *sepals*, collectively called the *calyx*. The calyx and corolla combined are referred to as the *perianth*. There may be leaflike *bracts* just below the sepals.

The male parts of the sexual organs are the *stamens*, and there may be one to several dozen in a flower. The stamens usually consist of hairlike *filaments* bearing the pollen-producing *anthers* on the outer ends. At the appropriate time, the anthers *dehisce* or split open and disgorge the male element, the numerous microscopic and usually yellow grains of pollen.

The female part of the flower is the *pistil*, consisting of the *ovary*, with one to numerous *ovules* or potential seeds, and extending from the ovary is the *style*, with the pollen-receptive portion, the *stigma*, on or near the tip. The pistil may be composed of one or more *carpels* or sections. Typically, the ovary, with its style and stigma, surrounded by the stamens, occupies the central area of the flower. Nectar usually is secreted at the base of the pistil, inside the corolla (fig. 2). Nectaries also may occur outside the corolla. These usually are referred to as extrafloral nectaries and do not contribute to pollination.

The ovules produce the seeds, and the ovary develops into the fruit. Usually one ovule must be fertilized for each seed that develops. If no seeds are produced, the fruit is unlikely to develop although a few fruits (certain cucumbers and citrus varieties) develop *parthenocarpically* (without being pollinated). If an insufficient number of seeds develop, the fruit is likely to be asymmetrical or otherwise not fully developed.

In general, the sooner pollination can occur after a flower opens, the greater the likelihood that fertilization of the ovules and seed development will occur. As time elapses, the pollen may be lost to insect foragers, wind, gravity, damage by heat, moisture, or drying out. Also, processes may set in that result in the shedding of the fruit.

Pollination and Fertilization

Usually, we think of pollination in the combined sense of transfer of pollen and set of fruit or seed. Actually, two sets of factors are involved: (1) transfer of viable pollen from the anther to a receptive stigma, and (2) sprouting of the pollen grain and growth of the pollen tube down the style into the ovary, and ultimately the union of male nuclei of the pollen grain with female germ cells in the ovule that results in seed development. Pollination is of no value without fertilization.

Sterility, Fertility, and Compatibility

Most flowers have both male and female functional parts. Some plants, however, such as asparagus (fig. 3), coriander (fig. 4), dill (fig. 5), or squash

---

[1] Deceased.

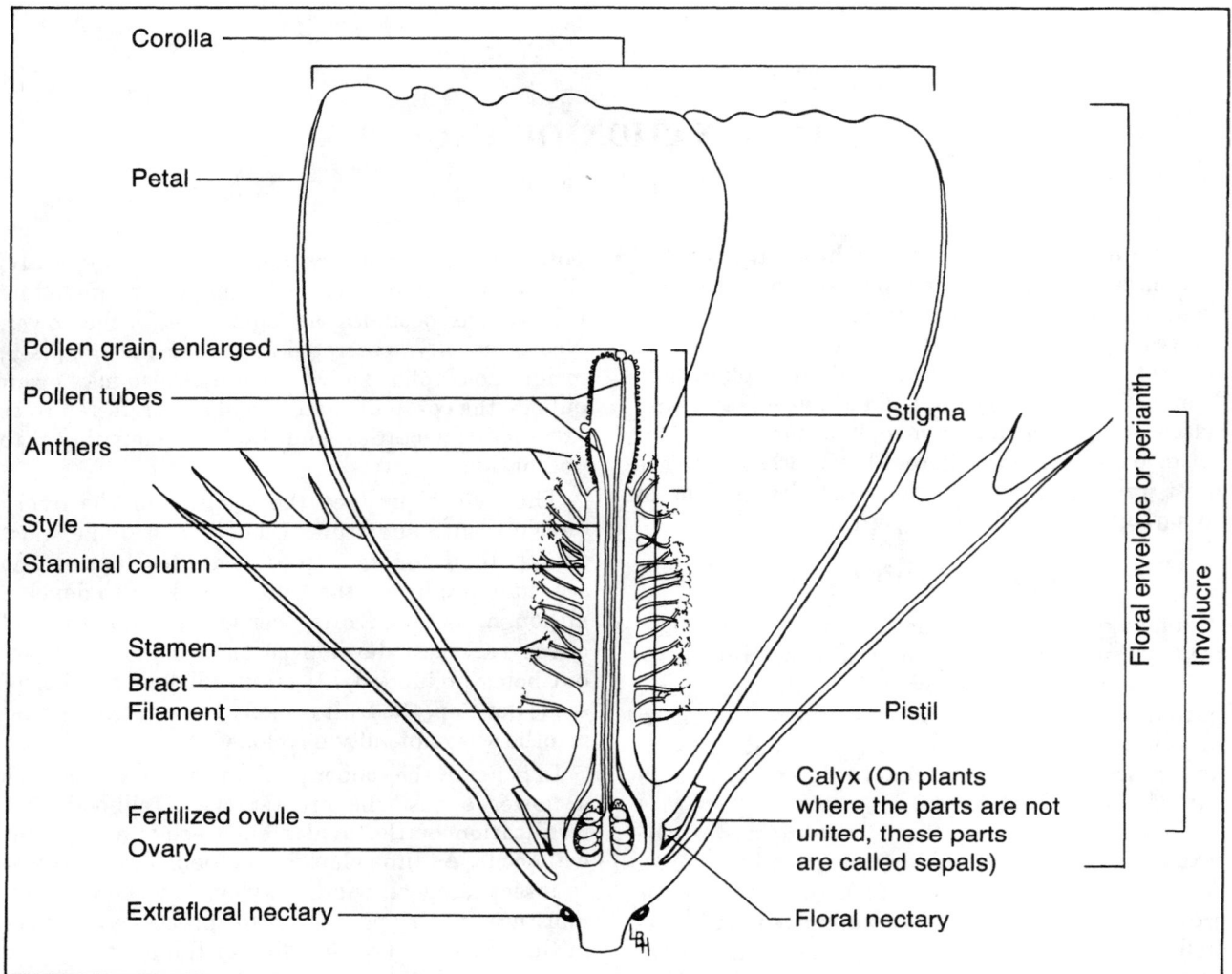

FIGURE 1.—Generalized longitudinal section of a cotton flower (*Gossypium* spp.), approximately × 2, showing nectaries, pollen-laden anthers, and growth of a pollen tube (further enlarged) down the style to the ovary and into an ovule.

(fig. 6) may have only *male* flowers, in which the ovaries are nonfunctional, or *female* flowers, in which the anthers are nonfunctional. In others, the stigma may not be receptive when the pollen is available within the flower (fig. 7). In such flowers, the pollen must be transferred from the male flower to the female. If the flower has both functional parts and is receptive to its own pollen, it is said to be *self-fertile*. If the flower is not fertilized by its own pollen, but is fertilized only when pollen comes from another plant or variety it is referred to as *self-sterile*. For example, both the 'Red Delicious' and 'Golden Delicious' apple are largely self-sterile, but when interplanted and cross-pollinated each will fertilize the other and

good production of fruit is obtained. They are *cross-compatible*. Varieties that will not cross-fertilize are said to be *cross-incompatible*.

A plant may be self-fertile but not *self-pollinating*. A pollinating agent may be necessary to transfer the pollen from the anthers to the stigma.

The avocado flower is an unusual example of the need for cross-pollination by bees (fig. 7). It opens twice on subsequent days. The first day the stigma is receptive to pollen but none is released by the anthers. After a few hours, the flower closes. The next day, it opens again when the stigma is no longer receptive to pollen, but the anthers release sticky clumps of pollen. Again the flower soon

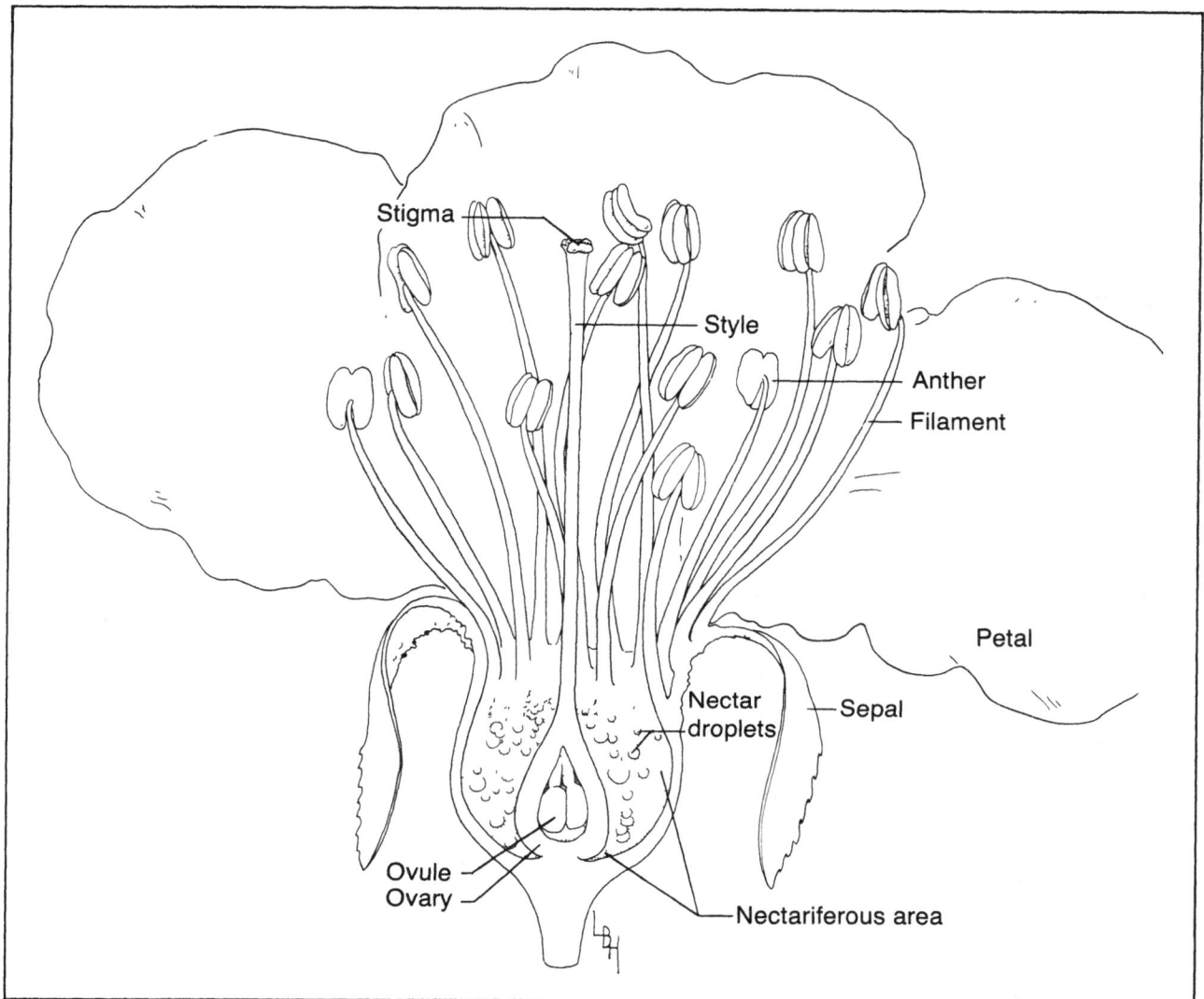

FIGURE 2.—Longitudinal section of a Bing cherry flower, approximately × 7, showing nectariferous area and nectar droplets.

closes, never to reopen; therefore, it cannot be self-pollinated.

Still more unusual is the characteristic of some varieties of avocado flowers to open for the first time in the morning of the first day and in the afternoon of the second day. In other varieties, the flowers open in the afternoon of the first day and in the morning of the following day. Only when growers interplant two such varieties, where pollen is always available when stigmas are receptive, and they provide bees to serve as the cross-pollinating agents, are they capable of harvesting the maximum set of fruit.

In addition to the volume of the crop produced through adequate pollination, another value lies in the effect of pollination on quality and efficiency of crop production. Inadequate pollination can result in reduced yields, delayed yield, and a high percentage of culls or inferior fruits. With ample pollination, growers may be able to set their blooms before frost can damage them, set their crop before insects attack, and harvest ahead of inclement weather. Earliness of set is an often overlooked but important phase in the crop economy.

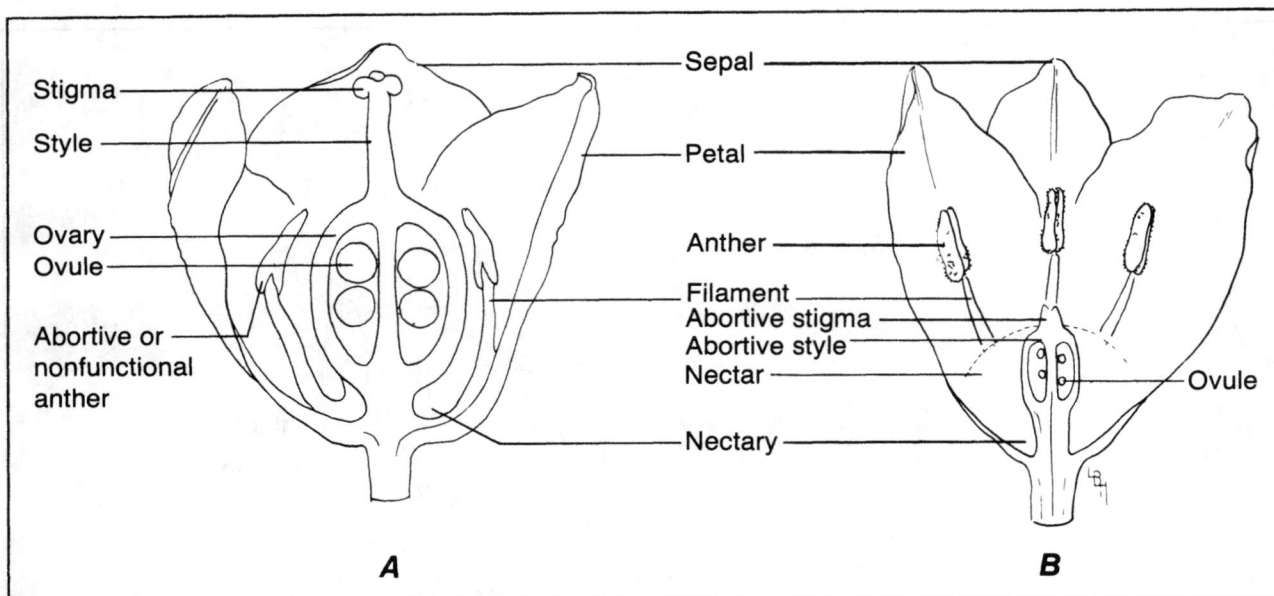

FIGURE 3.—Longitudinal section of asparagus flower, with fully developed ovary and stigma (A) and anthers (B), approximately × 17.

Hybrid Vigor and Bee Pollination

The value of pollination on the succeeding generation of crops also is frequently overlooked. The value of hybrid seed in not reflected until the next generation. Vigor of sprouting and emerging from the soil often is a vital factor in the plant's early survival. Other responses to hybrid vigor include earliness of development, plant health, and greater and more uniform production of fruit or seed.

When two unlike varieties are cross-bred, the offspring frequently is more robust in some characteristic than either parent. This strengthening effect is referred to as *hybrid vigor* (heterosis).

Years ago, scientists learned that the offspring of two inbred lines of corn was more productive than either parent. Then they learned that when this offspring was crossed with another such offspring of two other inbred lines, ever greater production was obtained. As an example, the offspring of variety A crossed with variety B, or A × B, crossed with C × D, is known as a 4-way cross. Presently, most of the corn produced in the United States is derived in this way, and is generally referred to as hybrid corn.

Corn is wind-pollinated; therefore, breeders plant certain rows of variety A between rows of variety B. Then, by removing the pollen-producing tassels of one variety, production of seed can only result from pollen carried by wind from the tassels of the other. This system works well on corn because the tassels can be deftly removed before pollen is released.

In most of our other crops, the male and female parts are in the same, usually small, flower. In some crops, however, breeders have developed methods of producing and maintaining *male-sterile* lines—selections that produce no pollen. Then alternate rows, or groups of rows, are planted to normal lines and others to male-sterile lines and all the fruit or seed obtained from the male-sterile line is hybrid. The hybrid may be superior to the parents in productiveness, uniformity, earliness, resistance to diseases or insects, or other factors.

Growers currently are producing hybrid onions, carrots, cucumbers, sunflowers, and several other crops. Research is under way on the production of hybrid cotton, soybeans, and alfalfa. All of these are insect-pollinated crops and will need to use colonies of bees or other pollinating insects in their production.

Pollinating Agents

Although this handbook principally concerns honey bees and beekeeping, pollinating agents other than honey bees should be mentioned and their relative value considered.

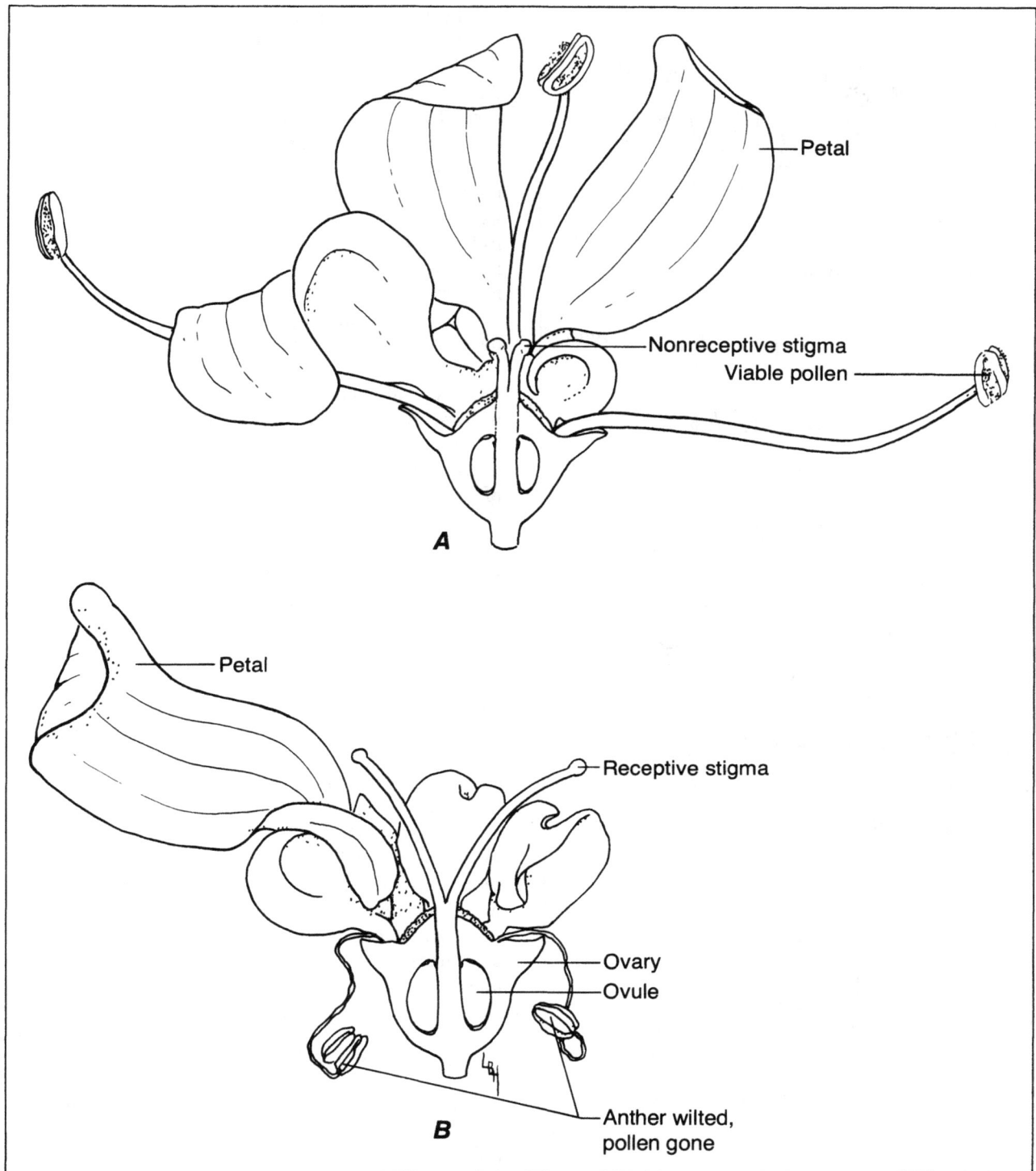

FIGURE 4.—Longitudinal section of corlander flower, approximately × 40: *A*, Staminate stage; *B*, pistillate stage.

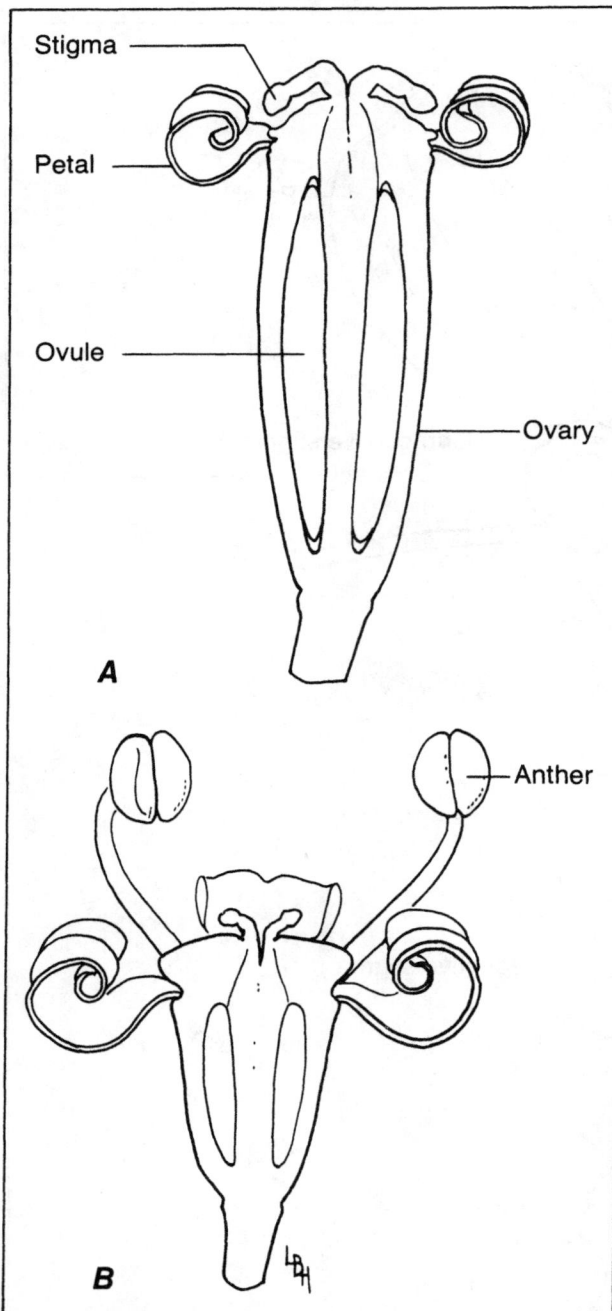

FIGURE 5.—Longitudinal section of dill flower, approximately × 20: *A*, Female, *B*, male.

Wind probably is the most important pollinating agent, insofar as it benefits our existence. Most of the forest trees, practically all the grasses and grains, with the exception of some that are completely self-pollinated, and many weeds are wind-pollinated.

The flowers of most wind-pollinated plants are either male or female. The male flowers produce an abundance of pollen to be carried by the wind. The female flowers usually have large stigmatic areas to receive the pollen. Corn is a good example of a wind-pollinated crop.

Birds of several different species feed upon nectar, pollen, or insects in some flowers and serve as pollinators. None is of significance in pollinating our cultivated crops. Their visits are confined largely to deep-throated, usually showy wild flowers.

Insects of many species visit flowers and pollinate them. These include bees, wasps, moths, butterflies, beetles, thrips, and midges. Bees are the most efficient and the only dependable pollinators, because they visit flowers methodically to collect nectar and pollen and do not destroy the flower or the plant in the process. Various species of bees, including the managed wild bees (see section on Management of Wild Bees), are highly efficient. An estimated 80 percent of our insect pollination is done by bees.

Honey Bees and Pollination

Modern agriculture has come to depend greatly upon honey bees to fulfill its pollination needs (fig. 8). This insect has several valuable qualifications for this role (fig. 9). Beekeepers maintain honey bees at a high population level in most agricultural areas of the United States for the honey and wax they produce. The colonies easily can be concentrated when and where needed to satisfy pollination requirements, and by using techniques developed for honey production their numbers can be increased in a relatively short time. The honey bee is adapted to many climates and can successfully revert to its original wild state in most parts of the world, quickly becoming part of the natural reservoir of pollinators.

Beekeeping in Relation to Pollination

When growers conclude that their crops must be insect-pollinated but do not have sufficient pollinators to do the job, they may decide to rent colonies of honey bees from a beekeeper. There may be a local contractor with whom growers can deal—one who acts as contact agent on behalf of a few neighboring beekeepers. Most of the bee rentals, however, are personal arrangements between growers and local beekeepers. Unfortu-

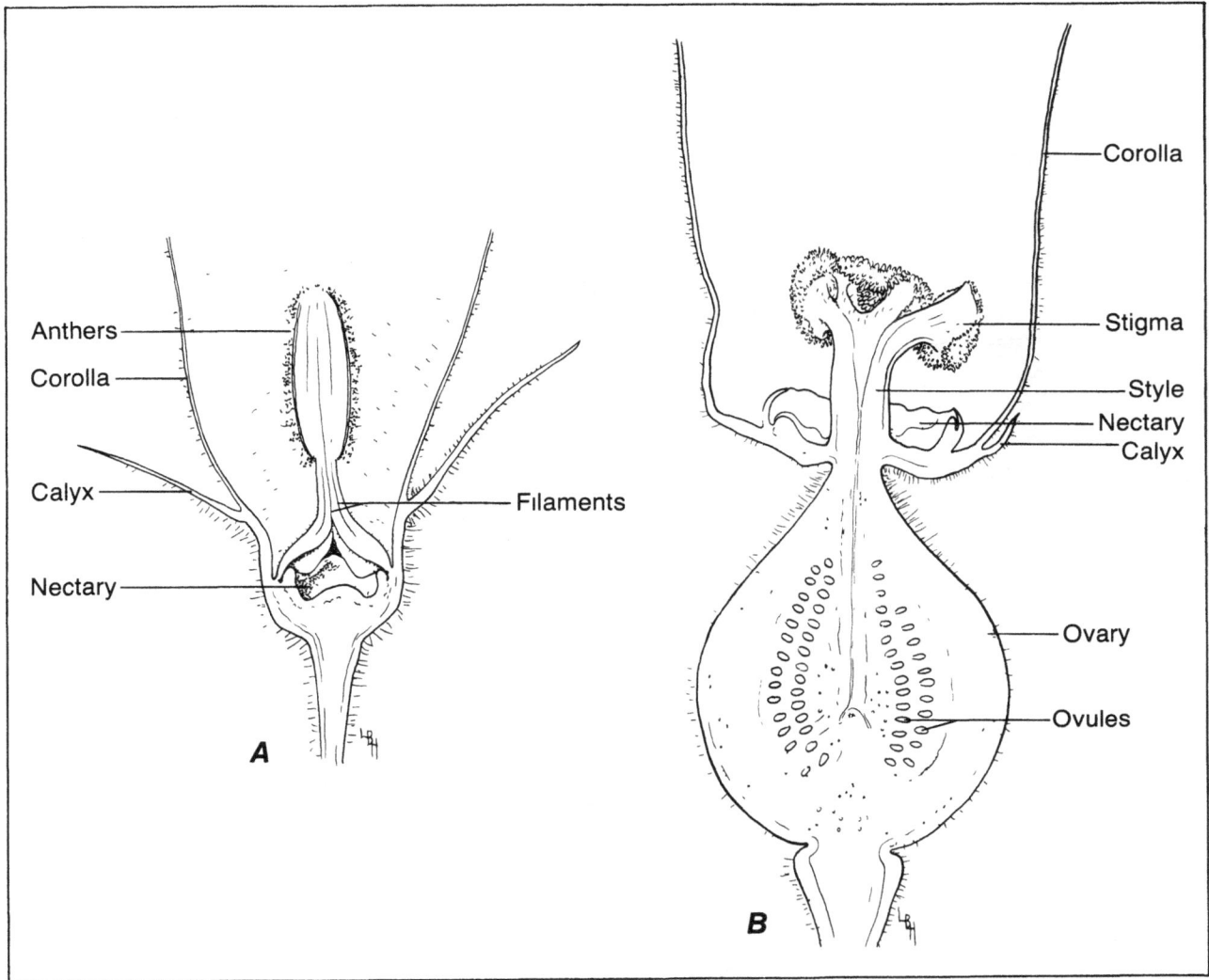

FIGURE 6.—Longitudinal section of reproductive portions of acorn squash flowers, approximately × 2: *A*, Staminate or male flower; *B*, pistillate or female flower.

nately, growers and beekeepers consider the rental arrangement from entirely different points of view.

Growers may consider only the fees they are paying and the potential value of the bees to their crops, along with having the beekeeper, the beekeeper's vehicles and crew, and the bees on their premises. Growers want to buy a service, pay for it, then go about their other duties.

Beekeepers consider the value of the bees to themselves before, during, and after the pollination contract is concluded. They weigh the advantage of the rental fee against the possibility of a reduced honey crop and the possibility of better forage than in their permanent location. (Beekeepers are always looking for better locations.) They also consider the constantly threatening adverse effect of pesticides, and the danger of damage to the bees and equipment in making the move. The bees usually are moved at night, and this is hard work with many chances for accidents on the roads or in the fields. When the colonies are moved from a location, another beekeeper may take it over. Exposure to bee diseases increases when colonies, particularly from many beekeepers, are concentrated in one area. And finally, problems associated with collecting the fee after the service is rendered frequently develop. These disadvantages discourage many beekeepers from renting bees to growers.

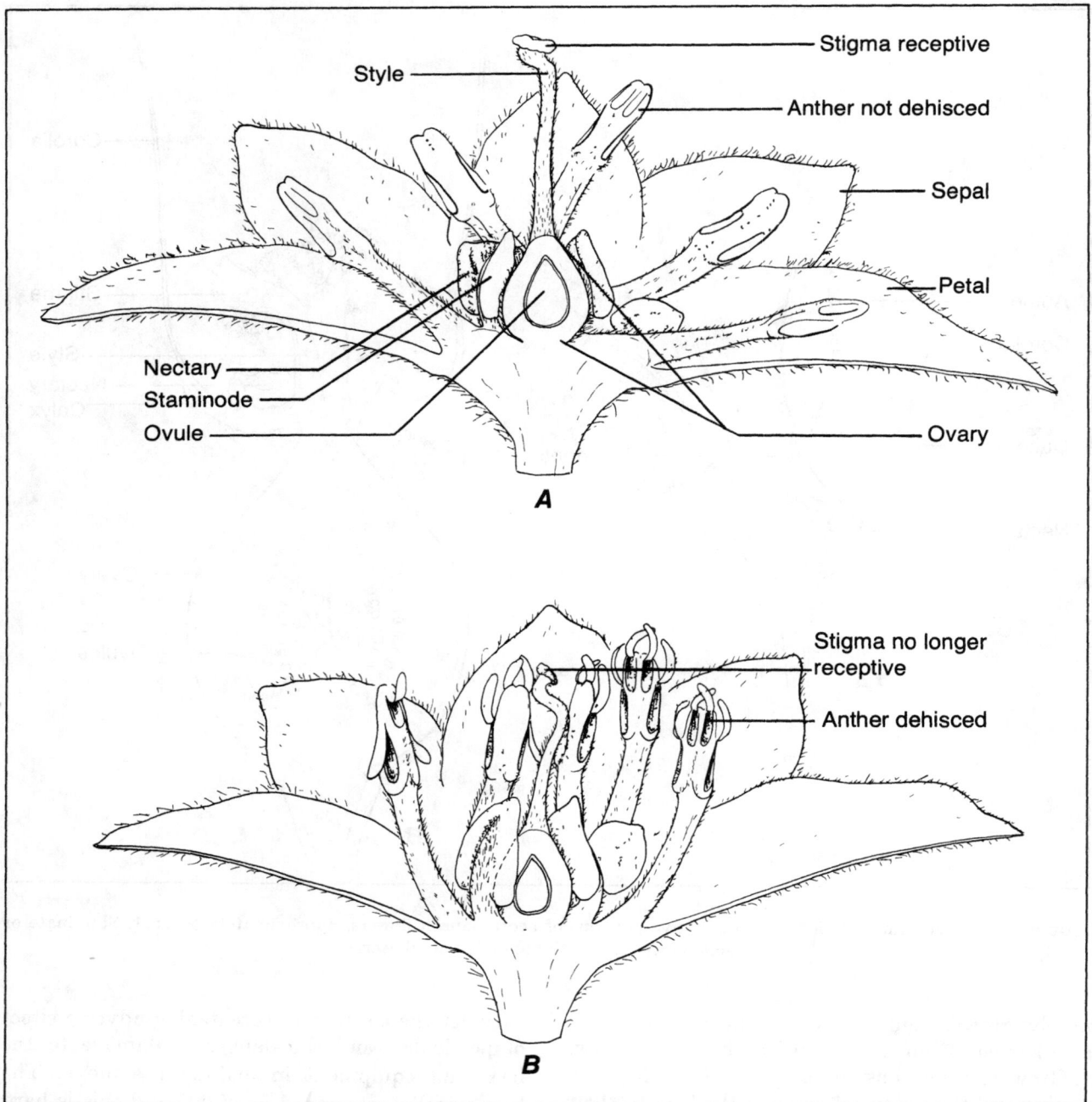

FIGURE 7.—Longitudinal section of 'Fuerte' avocado flower, approximately × 18: *A*, Stage 1: Stigma receptive, but stamens bent outward and anthers not dehisced; *B*, stage 2 the second day with stigma no longer receptive but stamens upright and anthers dehisced.

FIGURE 8.—Honey bee colonies in apple orchard.

FIGURE 9.—Intense flight activity at the hive entrances usually indicates that the colonies are populous or "strong."

Crops Dependent Upon or Benefited by Insect Pollination

The following crops are dependent upon or benefited by insect pollination. Unfortunately, the information on their pollination needs is scanty or based upon earlier popular varieties.

For some, insect dependence is absolute; for others, the benefit ranges from scant to great. For most of them, there is a great need for study of current varieties under different environmental conditions to determine the precise dependency on insect pollinators.

Fruit and nut crops	Vegetable crops	Forage crops [1]	Oilseed crops [1]	Other
Acerola	Artichoke [1]	Alfalfa	Cotton	Buckwheat [1]
Almond	Asparagus [1]	Alsike	Flax	Cacao
Apple	Balsam pear	clover.	Peanut	Cashew
Apricot	Broccoli [1]	Arrowleaf	Rape	Chicory [1]
Blackberry	Brussels	clover.	Soybean	Clove
Blueberry	sprouts. [1]	Ball	Safflower	Coffee
Chestnut	Cabbage [1]	clover.	Sunflower	Kola
Chinese	Cantaloupe	Berseem	Tung	Lupines [1]
gooseberry.	Cardoon [1]	clover.		Tea
Coconut	Carrot [1]	Black medic		Many garden
Crabapple	Casaba melon	Broadbean		flowers. [1]
Cranberry	Cauliflower [1]	Cicer		
Currant	Celeriac [1]	milkvetch.		
Feijoa	Celery [1]	Crimson		
Gooseberry	Chayote	clover.		
Grape (some kinds)	Chervil [1]	Crownvetch		
Grapefruit	Chive [1]	Kenaf		
Guava	Coriander [1]	Kidneyvetch		
Jujube	Cowpea	Kudzu		
Lemon	Crenshaw	Lespedeza		
Litchi	Cucumber	Mung bean		
Loquat	Dill [1]	Persian		
Macadamia	Eggplant	clover.		
Maney sapote	Endive [1]	Pigeon pea		
Mango	Fennel [1]	Red clover		
Nectarine	Honeydew	Rose clover		
Olive	Kale [1]	Sainfoin		
Orange (some	Leek [1]	Scarlet run-		
kinds).	Lima bean	ner bean.		
Papaw	Muskmelon	Strawberry		
Papaya	Mustard [1]	clover.		
Passion	Onion [1]	Sweetclover		
fruit.	Parsley [1]	Sweetvetch		
Peach	Parsnip [1]	Trefoil		
Pear	Pepper	Vetch		
Persimmon	Persian	White clover		
Pomegranate	melon.	Zigzag clover		
Plum	Pimento			
Prune	Pumpkin			
Quince	Radish [1]			
Raspberry	Rutabaga [1]			
Tangelo	Squash			
Tangerine	Turnip [1]			
Temple orange	Vegetable			
	sponge.			
	Welsh			
	onion. [1]			

[1] For production of seed.

DISEASES AND PESTS OF HONEY BEES

By H. Shimanuki[1]

The first bee laws in the United States were enacted in 1883 to establish methods for control of bee diseases. Today, 49 States provide apiary inspection services for disease abatement. Bee diseases cause considerable expense to the States for the cost of maintaining apiary inspection service, as well as considerable losses to the beekeepers for the cost of colonies damaged or destroyed and for the drugs fed to prevent bee diseases. In addition, far greater losses result from

reduction in honey and beeswax production and insufficient bees for pollination. It is apparent, therefore, that both beginning and advanced beekeepers should learn to recognize and control bee diseases.

Brood Diseases

The most common brood diseases found in the United States are American and European foulbrood, sacbrood, and chalk brood. A guide for diagnosing brood diseases of honey bees is given in table 1.

---

[1] Laboratory chief, Science and Education Administration, Bioenvironmental Bee Laboratory, Beltsville, Md. 20705.

TABLE 1.—*Comparison of symptoms of various brood diseases of honey bees*

Symptom	American foulbrood	European foulbrood	Sacbrood	Chalk brood
Appearance of brood comb.	Sealed brood. Discolored, sunken, or punctured cappings.	Unsealed brood. Some sealed brood in advanced cases with discolored, sunken or punctured cappings.	Sealed brood. Scattered cells with punctured cappings, often with two holes.	Sealed and unsealed brood.
Age of dead brood.	Usually older sealed larvae or young pupae. Lying lengthwise in cells.	Usually young unsea'ed larvae; occasionally older sealed larvae. Typically in coiled stage.	Usually older sealed larvae; occasionally young unsealed larvae. Lengthwise in cells.	Usually older larvae. Lengthwise in cells.
Color of dead brood.	Dull white, becoming light brown, coffee brown to dark brown, or almost black.	Dull white, becoming yellowish white to brown, dark brown, or almost black.	Grayish or straw-colored becoming brown, grayish black, or black; head end darker.	Chalk white. Sometimes mottled with black spots.
Consistency of dead brood.	Soft, becoming sticky to ropy.	Watery; rarely sticky or ropy. Granular.	Watery and granular; tough skin forms a sac.	Pasty.
Odor of dead brood.	Slightly to pronounced putrid odor.	Slightly to penetratingly sour.	None to slightly sour.	Yeastlike, nonobjectionable.
Scale characteristics.	Lies uniformly flat on lower side of cell. Adheres tightly to cell wall. Fine, threadlike tongue of dead pupae may be present. Head lies flat. Black in color.	Usually twisted in cell. Does not adhere tightly to cell wall. Rubbery. Black in color.	Head prominently curled towards center of cell. Does not adhere tightly to cell wall. Rough texture. Brittle. Black in color.	Mummified. Does not adhere to cell wall. Brittle. Usually chalky white in color. Sometimes black fruiting bodies are present.

American Foulbrood

American foulbrood disease occurs throughout the world where honey bees are kept. About 3 percent of all colonies inspected in the United States are found to be infected.

Bacillus larvae White, the causative organism of American foulbrood disease, is a spore-forming bacterium which produces over a billion spores in each infected larva. Only spores are capable of inciting the disease. The spores are extremely resistant to heat and chemical agents. Worker, drone, and queen larvae are susceptible to the disease. Under natural conditions, infected queen and drone larvae are rarely seen.

A severely infected American foulbrood comb has a mottled appearance due to a mixture of healthy capped brood, cells containing the remains of diseased larvae, and empty cells. The cappings of cells containing disease appear moist and darkened. The convex cappings found on cells of diseased larvae become concave as the disease progresses. Another symptom commonly associated with the disease is the punctured capping. Larvae are susceptible to American foulbrood only when they are less than 3 days old. A healthy larva has a glistening, pearly white appearance. Normally it begins development curled on the base of the cell. As it grows, it elongates to the full length of the cell. It is in the elongated position that the larva or pupa dies. As the infection progresses, the larva or pupa changes to creamy brown and eventually becomes dark brown. The remains become ropy and can be drawn out as threads of an inch or more. A very unpleasant, foul odor develops at this stage. The odor resembles that of animal glues that are rarely used.

The remains of diseased brood finally dry down to form scales that adhere strongly to the lower sides of the cells. If death occurs in the pupal stage, the mouth parts may adhere as a fine thread to the upper side of the cell. This is a positive symptom of American foulbrood disease (figs. 1 and 2).

The infection can be transmitted to a larva from nurse bees or from spores remaining in the bottom of the brood cell. Exchanging combs containing remains of diseased larvae or honey, or both, laden with spores of *B. larvae* is the most effective way to spread the disease from colony to colony. Early detection of the disease is helpful in preventing further spread. A colony that is weakened by American foulbrood may be robbed, and the robber bees inadvertently carry honey containing spores of *B. larvae* to healthy colonies.

European Foulbrood

In some areas, European foulbrood is a more serious threat to beekeepers than American foulbrood. This disease is serious because it occurs most frequently at the time that colonies are building their peak populations.

The cause of this disease is *Streptococcus pluton* White, a nonspore-forming bacterium. Other bacteria commonly associated with the disease are *Bacillus alvei* Chesire & Cheyne and *Bacterium eurydice* White.

Superficial examination of diseased combs shows the same mottled effect and puncturing as seen in American foulbrood. Death usually occurs in the larval stage. Worker, drone, and queen larvae are equally susceptible to European foulbrood.

Larvae that die from European foulbrood are found in various positions. Some are in a curled stage and others elongated. The normal pearly white appearance of a healthy larva changes to a dull white, then yellow and finally brown. Ropiness and sour odor are caused by the secondary organisms associated with the disease. The elasticity of the ropy material is less than that associated with American foulbrood. The tracheae appear as fine silvery tubes immediately below the skin, especially as the larvae turn brown. This symptom is highly characteristic of European foulbrood. Loosely adhering scales also differentiate this disease from American foulbrood (fig. 3).

European foulbrood can be transmitted by contaminated food stores, and equipment. The disease usually is most serious in the spring and clears up during the summer when nectar and pollen are abundant. However, outbreaks of European foulbrood in the late summer are not unusual.

Chalk Brood

Chalk brood disease was not found in this country until 1968. Since that time, the disease has spread throughout the United States and Canada. No accurate figures are available on losses attributed to this disease. Chalk brood appears primarily in the spring, although outbreaks in the summer and fall can occur.

Ascosphaera apis (Massen ex Claussen) Olive and Spiltoir, a fungus, is the cause of chalk brood disease. Larvae, 3 to 4 days old, are most sus-

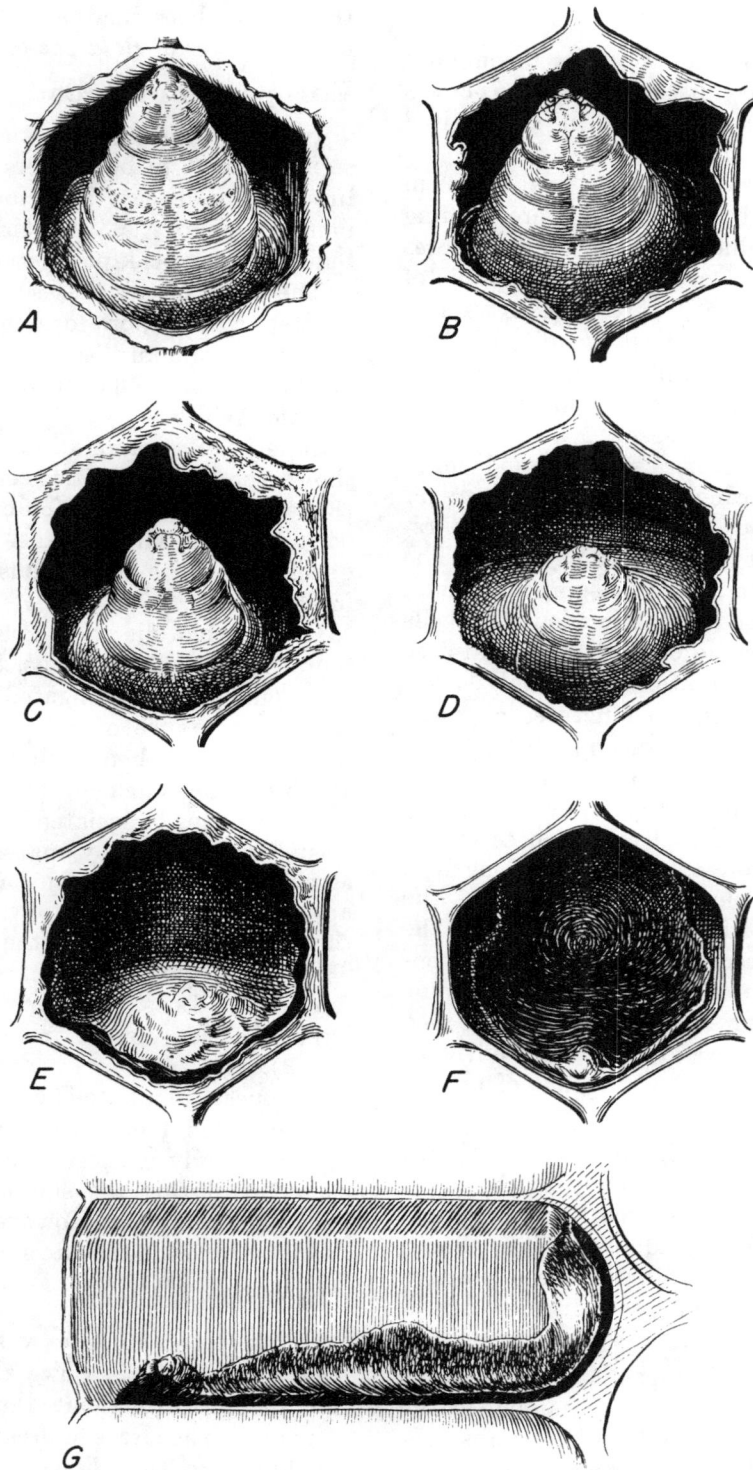

DN-2219

FIGURE 1.—Honey bee larvae killed by American foulbrood, as seen in cells: *A*, Healthy larva at age when most of brood dies of American foulbrood; *B–F*, dead larvae in progressive stages of decomposition (remains shown in *F* are scale); *G*, longitudinal view of scale.

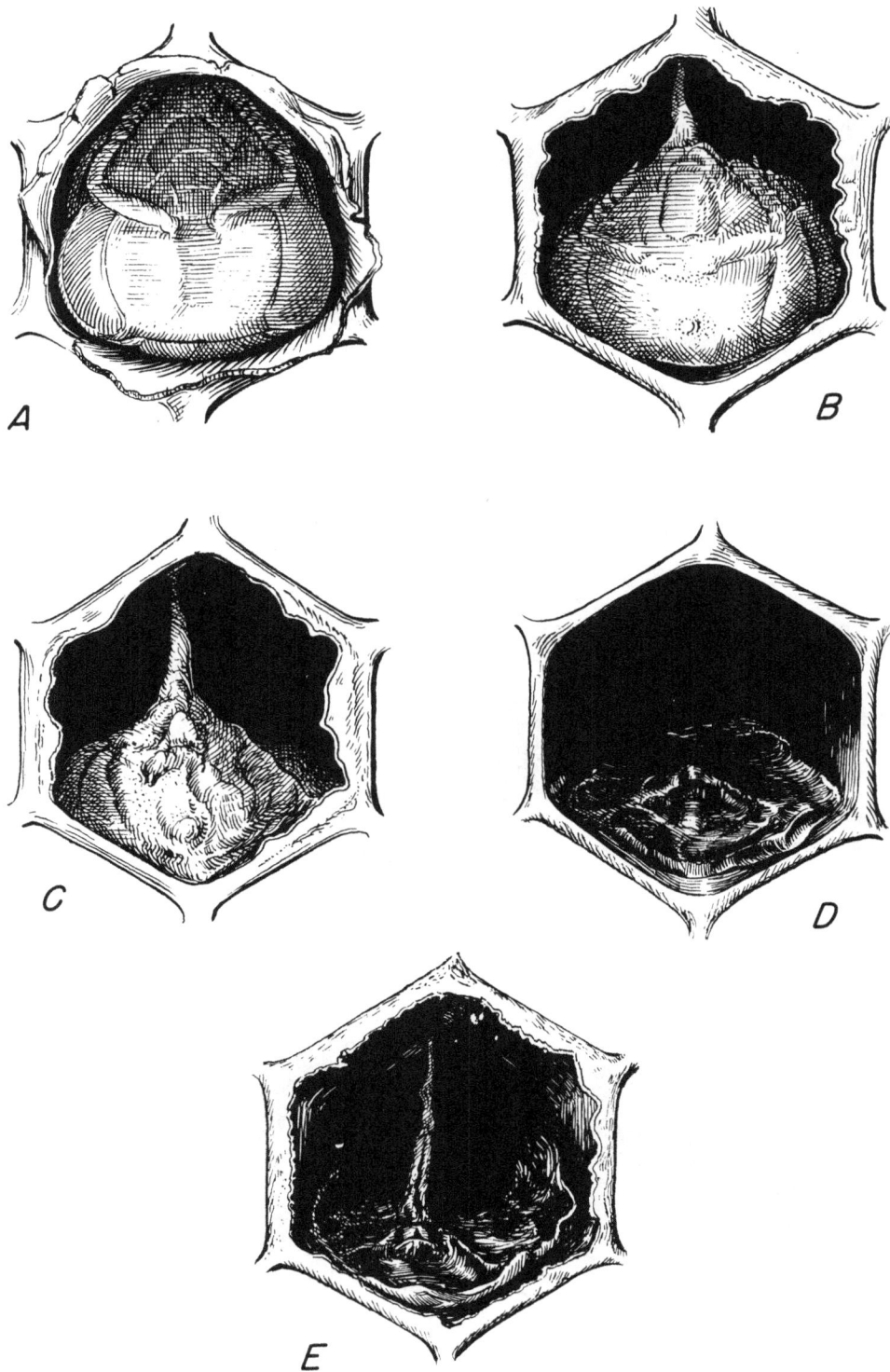

DN-2220

FIGURE 2.—Honey bee pupae killed by American foulbrood, as seen in cells: *A–C*, Heads of pupae in progressive stages of melting down and decay; *D–E*, scales formed from drying of dead pupae. In *B*, *C*, and *E*, tongue is shown adhering to roof of cell.

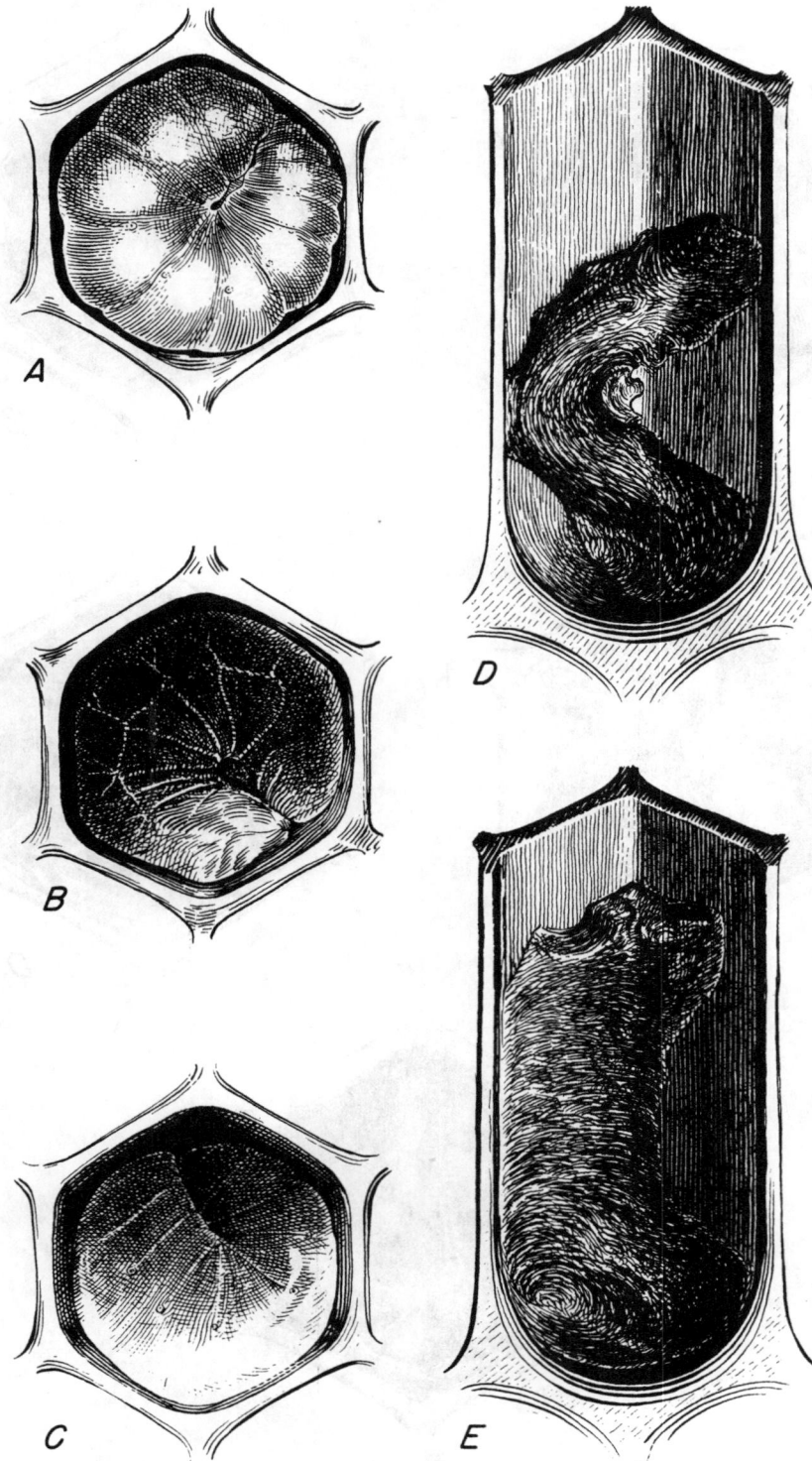

DN-2218

FIGURE 3.—Honey bee larvae killed by European foulbrood, as seen in cells: *A*, Healthy larva at earliest age when brood dies of European foulbrood; *B*, scale formed by dried-down larva; *C*, one position of sick larvae just before death; *D–E*, longitudinal views of scales from larvae that were in lengthwise position before death.

ceptible to the fungus, especially if they are chilled after ingesting spores of *A. apis*. Worker, drone, and queen larvae all are susceptible to chalk brood disease.

Infected larvae become permeated with the mycelia of the fungus, leading to their death. Eventually, the mycelia-filled larvae dry up to form the typical hard white mummies characteristic of chalk brood disease. Diseased larvae are stretched out in their cells and also can be mottled or completely gray or black. This color variation is due to the presence or absence of the black fruiting bodies that are formed on the outside of the larvae.

The disease can be detected by examining the combs, the entrance, and bottom boards of the hives for the presence of the mummies. The mummies do not stick to the cells and are easily removed by nurse bees. If colonies have pollen traps, the mummies frequently are found in the traps and are a source of infection in trapped pollen. The mummies have a faint yeast-like odor.

Chalk brood disease can be transmitted by adult bees and equipment contaminated with spores of *A. apis*. The disease appears to clear up spontaneously and may reappear later in the season or the following year. The disease rarely destroys a colony but can reduce the population of bees and consequently affect the honey yield.

Sacbrood

Death of a colony by sacbrood is rare. Because of the similarity to other diseases, however, the beekeeper should learn to distinguish sacbrood from the more serious diseases. The etiologic agent in sacbrood is a virus.

Larvae die of sacbrood in capped cells in the elongated position. As the disease progresses, the larval skin forms a sac, which separates from the prepupal skin. Between these two layers of skin is an accumulation of fluid. The outer skin toughens and, as a result, the larva can be picked up in its entirety without the release of the fluid.

The larva changes from pearly white to off-white, then brown, and finally almost black. The head of the larva usually curls up from the cell floor (fig. 4). A loosely adhering scale is formed from the larval remains. It has the appearance of a foulbrood scale but no odor, and is free of bacteria.

Like European foulbrood, sacbrood is most commonly found in the spring. No chemotherapeutic agent is effective against sacbrood. Re-

queening may be helpful, but most colonies appear to recover spontaneously from the disease.

Other Blood Diseases

Aspergillus flavus Link, a fungus, usually is isolated from bees that have stonebrood. This disease is unusual in that it infects both brood and adults. Bees dying from this disease form mummies. The fruiting bodies of the fungus make the infected bee appear yellowish-green or brown.

Purple brood is a nutritional disease of larvae and pupae. It is believed to be caused by nectar or pollen from *Cyrilla racemiflora* L., also called southern leatherwood, black titi, red titi, summer titi, and he-huckleberry. This problem exists only in the Southern States, where southern leatherwood is found. Diseased larvae and pupae are purple.

Other conditions that mimic contagious diseases are chilled and starved brood. Chilled brood is caused by lack of sufficient bees to keep the brood area warm. Consequently, chilled brood usually is found at the outer edge of the brood nest. Brood may have the appearance of European foulbrood but is readily removed by nurses bees as the brood pattern expands. Starved brood generally is caused by insufficient nectar or honey. At times, wax moth damage to developing bees may cause them to appear diseased.

Adult Diseases

Nosema Disease

Nosema disease is the most widespread of all bee diseases. It was found in over 60 percent of the apiaries sampled in the United States. This disease is caused by the protozoan, *Nosema apis* Zander. Nosema disease reduces the life expectancy of adult bees. It can cause queen supersedure and reduce the honey production of infected colonies.

The disease cycle is initiated by adult bees that ingest spores of *N. apis*, which germinate and multiply in the epithelial cells of the ventriculus. In addition to affecting the digestive process, the hypopharyngeal glands of infected worker bees and the ovaries of infected queens become atrophied. The disease is found in workers, queens, and drones.

No external symptoms may be visible in bees or in colonies infected with *N. apis*. Some of the infected ventriculi may become distended and white. Nosema disease is diagnosed by examining for the presence of *N. apis* spores. However, the absence of spores does not ensure freedom from

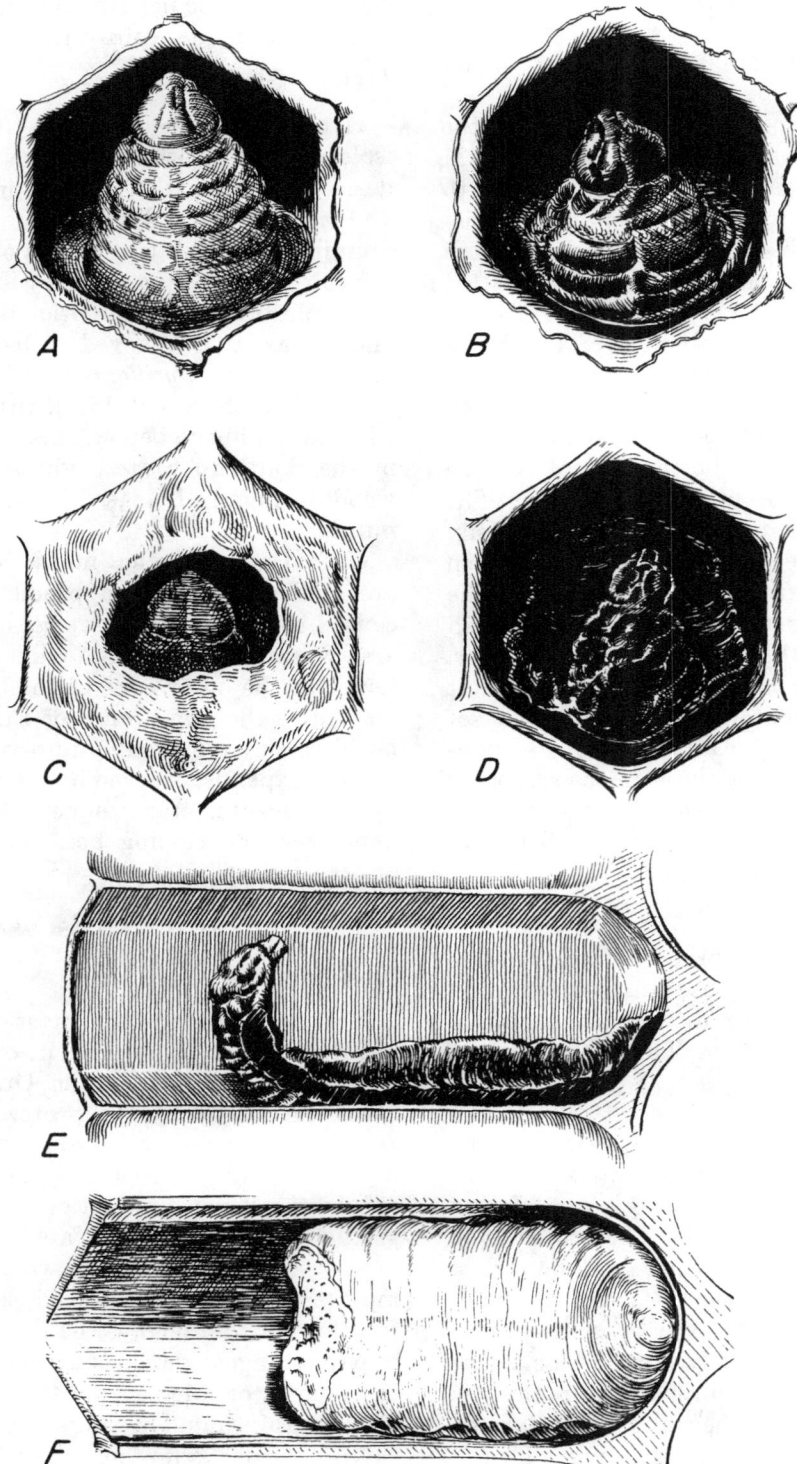

DN-2217

FIGURE 4.—Honey bee larvae killed by sacbrood, as seen in cells: *A–B*, larvae in different stages of decomposition; *C*, erect head of dead larva showing through opening made by bees in capping; *D–E*, views of scale (note how head remains erect); *F*, remains of larva, head of which has been gnawed away by bees.

nosema disease, since other life stages of the protozoan may be present. The disease has an annual cycle which results in maximum numbers of spores in the spring. Spore numbers decline in the summer, and in some cases, a small peak in the fall may be visible.

Virus Bee Paralysis

Bee paralysis is caused by several different viruses, but some nectars and pollens also may induce similar symptoms. Chronic bee paralysis and hairless black syndrome are caused by the same virus. Acute bee paralysis, caused by another virus, kills bees more quickly than the chronic virus.

Affected bees quiver and cannot fly. Frequently, they appear greasy and shiny with no hair on their thorax. The disease is transmitted to healthy bees when they attack diseased bees or when food is exchanged between healthy and diseased bees.

Workers, drones, and queens are susceptible to chronic bee paralysis. It appears that susceptibility to the disease is inherited from the queen. Consequently, requeening of colonies with the disease may rid the colony of all symptoms.

Septicemia

Septicemia is a bacterial disease of adult honey bees that is rarely encountered; it is caused by *Pseudomonas apiseptica* Burnside.

The bacteria, by some unknown method, make their way to the hemolymph, multiply rapidly, and ultimately cause the death of the host.

Bees that die from septicemia appear to have no connective tissues and dismember easily. The legs, wings, head, thorax, and abdomen separate—even by the slightest handling. Hemolymph of infected bees also may be milky white in color. The isolation and identification of *Ps. apiseptica* bacteria from the hemolymph may be necessary for verification.

Mite Diseases

The Asiasitic mites, *Varroa jacobsoni* Oudemans and *Tropilaelaps clareae* Baker and Delfinado, both affect larvae and pupae of the honey bees. *Varroa jacobsoni* has been found in South America but is not present in North America. However, *T. clareae* has not been found on bees outside Southeast Asia.

The mature female *V. jacobsoni* mite attaches itself to bees and can be transmitted from colony to colony by robbing and drifting bees. The female mite lays her eggs in cells containing larvae just before being sealed. After the eggs hatch, the nymphs feed on the developing larvae or pupae, causing malformation and sometimes death of the host.

Acarine Disease

Acarapis woodi Rennie, the causative agent of acarine disease, has not been found in North America. The Honeybee Act enacted in August 1922 and amended several times since prohibits the importation of all live stages of the honey bees and was written principally to prevent the entry of *A. woodi* into North America.

The mite enters its host via the spiracles and spends most of its life in the thoracic tracheae. At maturity, the female mite emerges from the tracheae in search of a new host. No one symptom characterizes this disease; an affected bee could have disjointed wings and be unable to fly or have a distended abdomen, or both. Positive diagnosis of this disease can be made by microscopic examination of the tracheae for discoloration (black spots) and the presence of eggs, nymphs, and adult stages of the mite.

External Mites

External mites frequently are present on bees in the United States. These mites closely resemble *Acarapis woodi* but cause no apparent harm to the bees. A microscope is necessary to differentiate external mites from *A. woodi*.

Other Adult Diseases

Amoeba disease and gregarines are protozoans sometimes found in honey bees in North America. Neither disease seems to be of economic importance.

Amoeba disease is caused by *Malpighamoeba (Vahlkampfia) mellificae* Prell. This disease sometimes is found together with nosema disease, and the combination may be more serious than either disease alone. The cysts of the amoeba are transmitted by the excreta of bees, and the infection is localized in the Malpighian tubules. It is believed that the amoeba interferes with the function of the Malpighian tubules, which ultimately leads to the death of the bee. Diagnosis for this disease is made by microscopic examination of the Malpighian tubules for the presence of amoeba cysts.

Gregarines are found in the digestive tract of adult bees. No pathological significance, however, has been attached to these protozoans. Several different genera of gregarines are found in the United States.

Wax Moths

The most serious pest to honey bee colonies is the greater wax moth, *Galleria mellonella* L. In addition to the greater wax moth, comb damage is caused by the lesser wax moth, *Achroia grisella* F. and the Mediterranean flour moth, *Anagasta kuehniella* Zeller.

Damage by the greater wax moth is severest in the Southern United States because of the long warm season and high temperatures. The wax moth distribution, however, includes all areas where honey bees are kept. It is not a threat to normal colonies and cannot kill a colony, but weakened colonies are invaded and unused combs destroyed.

Female wax moths lay their eggs on combs or in cracks between the wooden parts of the hives (fig. 5A). After egg hatch, the larvae (fig. 5B) feed on the wax combs, obtaining nourishment from the cast-off honey bee pupal skins, pollen, and other impurities found in the combs. For this reason, darkened combs are more likely to be infested than light combs or foundation.

The fully grown larva spins its own cocoon, which usually is attached to wooden parts inside the hive—such as the inner cover, hive body, and frame. In colder climates, the greater wax moth overwinters as a pupa. In warmer areas, adults emerge all year. The adult female (fig. 5C) is about ¾-inch long and 1 to 1¼ inches wide from wingtip to wingtip. Within 4 to 10 days after emergence, the female begins to lay eggs. She lays about 300 eggs in her lifetime, which usually is somewhat less than 3 weeks.

Combs are most often destroyed by the wax moth when stored in dark, warm, and poorly ventilated rooms (fig. 6). However, there can be considerable damage to combs even while in use, especially in hives where the population of adult bees is too small to protect all the combs.

Paradichlorobenzene and ethylene dibromide have been used in the past for the control of wax moths. Other control measures include carbon dioxide, heat, or cold treatments.

Larvae of the Mediterranean flour moth and the lesser wax moth also can cause damage to combs in storage and in the hives. The damage caused by these insects is quite similar, and it is necessary to identify the insects to be certain which is causing the problem. The same control methods work for all three insects.

Other Insect Pests

In some areas of the United States, termites may damage the wooden parts of the hive. The termites do not affect the bees directly. Although ants can be found in hives, they rarely cause any problem. Ants are more of a nuisance to the beekeepers than to the bees. Other insects such as certain wasps and robber flies also prey on honey bees but are considered of no economic importance in the United States.

Braula coeca Nitzsch, the "bee louse," is found primarily in the Mid-Atlantic States. The bee louse actually is a wingless fly and not a louse. This insect can be found on drones, workers, and queens. The destructive stage of the insect is the larva, which burrows under the cappings of honeycombs and ruins what would be good comb honey. No apparent damage is attributed to the adult bee louse, which spends its life on the bodies of workers and queens.

Enemies of Bees

Bears cause severe damage to the hives as they feed on the honey, adult bees, and brood. Electric fences and bear platforms have been used successfully to prevent bear damage.

Skunks, birds, toads, and frogs also feed on adult bees. Skunks can cause serious damage, as they consume large numbers of bees and thereby deplete populations to critical levels.

In the winter, mice can enter hives without entrance reducers. Although mice do not feed on bees, they chew on the combs and frames to construct nests in a warm, secluded area of the hive where they are not disturbed by the clustered bees.

Control of Bee Diseases

The U.S. Food and Drug Administration approved labeling for oxytetracycline and fumagillin as aids in the control and prevention of bee diseases. Oxytetracycline is effective in the control of both American and European foulbrood disease. Fumagillin is used to control nosema disease. Users should read the container label—it has specific instructions for the use of these materials. These drugs can be used subject to State laws and regulations in the manner specified but should

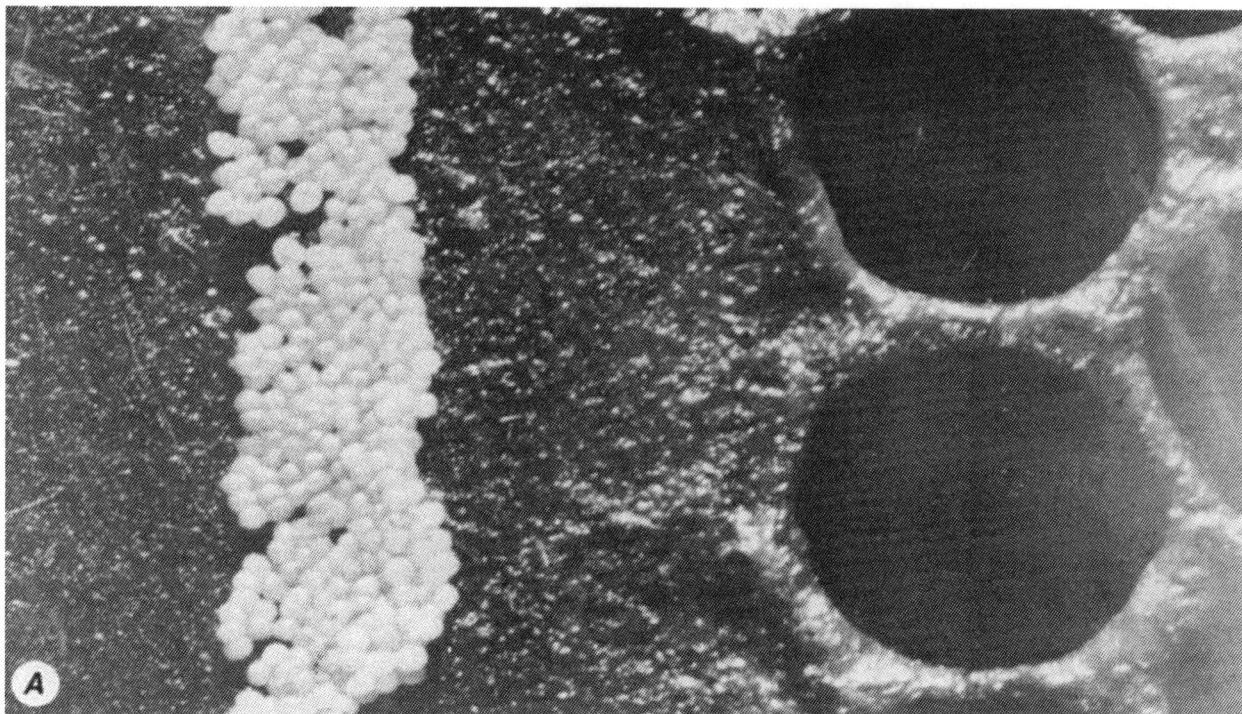

FIGURE 5.—Greater wax moth: *A*, eggs; *B*, larvae (left, dorsal; right, lateral); *C*, adult.

FIGURE 6.—Brood comb infested with greater wax moth larvae.

never be used at a time or in such a way that would result in contamination of the marketable honey. Some States require that the bees, contaminated combs, and honey from infected colonies be destroyed by burning. Beekeepers should consult their local apiary inspector for instructions on the disposal of diseased hives and the use of drugs.

Sending Samples for Laboratory Examination

If only a small amount of the brood or a few bees are affected or if the symptoms are unusual, a definite diagnosis in the apiary is sometimes difficult. Examination by laboratory methods is then necessary. Sometimes laboratory verifications of diagnoses made in the apiary also are desirable.

Diagnosis of disease in the laboratory is a service made available to beekeepers and State apiary inspectors by the U.S. Department of Agriculture.

A sample of brood comb for laboratory examination should be 4 or 5 inches square and contain as much of the dead brood as possible. *No honey should be present*, and the comb should not be crushed. A sample of adult bees should consist of at least 200 sick or recently dead bees.

Mail the samples in a wooden or strong cardboard box. Do not use a tin, glass, or plastic container, and do not wrap the comb or bees in waxed paper or aluminum foil. Send all samples to the U.S. Department of Agriculture, Science and Education Administration, Bioenvironmental Bee Laboratory, Building 476, BARC–E, Beltsville, Md. 20705. Your name and address should be plainly written on the box. If the sample is forwarded by an inspector, his or her name and address also should appear on the box.

PESTICIDES[1] AND HONEY BEE MORTALITY

By William T. Wilson, Philip E. Sonnet, and Adair Stoner[2]

Introduction

In a world where people expect and demand more food and fiber each year, all branches of agriculture must continuously adapt and improve to meet this challenge. Farmers now find it essential to annually increase efficiency and production to remain in business and to show a profit.

Many major agricultural changes took place in the 1950's, shortly after World War II, when tractors replaced horses, chemical fertilizers replaced organic manure, aerial application of pesticides became commonplace, and farmers became increasingly conscious of business costs. At the same time, many farmers were encouraged to devote large acreages to the cultivation of a single crop, which necessitated the utilization of large quantities of synthetic fertilizers and pesticides to nourish and protect that crop. Consumers also came to expect all market fruits and vegetables to be completely free from insects and insect damage.

Thus, many growers found it advantageous to apply more and more pesticides each year. Unfortunately, some aspects of this agricultural modernization were not beneficial for beekeepers, whose needs were either frequently forgotten or ignored. Consequently, many honey bees were killed. To compensate, many commercial beekeepers had to keep larger numbers of honey bee colonies in a variety of locations to make up for losses from pesticides and to meet rising operating expenses. This interaction between the needs of crop farmers and the needs of beekeepers, coupled with frequent widespread application of the "newer" insecticides, such as parathion, proved devastating to thousands of colonies.

Not only do insecticides create problems for beekeepers, but herbicides have adverse effects as well. For example, farmers employ herbicides in the practice of *clean cultivation*, which destroys nearly all weeds along fencelines, irrigation ditches, and on wasteland. Under these circumstances, fewer nectar- or pollen-producing plants are left in farming regions, except for those plants under cultivation and consequently under insecticide treatment. The shift by farmers in some areas from planting vast acreages of alfalfa and clover to corn had a dramatic effect on the bee industry. Corn affords the bees no nectar, but many bees are killed while collecting pollen from insecticide-treated corn.

Another problem that has created difficulties for both crop farmers and beekeepers is the social and economic pressure to produce more food on fewer acres, since much fertile land has been taken over by shopping centers, housing projects, highways, and other projects designed to accommodate masses of people. Consequently, many beekeepers have been forced into areas of intensified agriculture where pesticide exposure is greatest. If pesticide application becomes too intense, the beekeeper again is excluded from large acreages; thus honey yields decrease.

On the brighter side, many farmers growing such crops as almonds, apples, cranberries, seed alfalfa, citrus fruits, and other cash crops have learned that pollination by honey bees and other insects is absolutely essential for maximum crop production. This dependency has resulted in a better understanding between crop producers and beekeepers. Unfortunately, however, the bee-mortality problem has been solved only partially. In other areas, very large numbers of honey bees still are killed each year by insecticides and other agricultural chemicals.

To illustrate the magnitude of the pesticide problem for bees, the quantity of pesticides pro-

[1] *Pesticides* include many different chemicals used mainly in agriculture to control pest plants and animals. This general category includes herbicides, fungicides, nematicides, miticides, rodenticides, insecticides, and inhibitory plant and animal growth regulators such as hormones. Insecticides are used specifically to control insect pests and are the largest group of pesticide chemicals.

[2] Location leader, research chemist, and research entomologist, respectively, Science and Education Administration, Bee Disease Laboratory, Laramie, Wyo. 82071.

duced and sold in the United States has increased every year since 1957, except for 1969 and 1970, when there were slight decreases. Production of synthetic organic pesticides in the United States in 1974 amounted to more than 1.4 billion pounds (709,000 tons). Of the total production, 650 million pounds were insecticides. Even after taking imports and exports into account, slightly more than half the production, 400,000 tons, of pesticide was applied in the United States (Fowler and Mahan *1976*). Currently, there are more than 50 insecticides in common use with moderate to high toxicity to honey bees (Atkins *1977*).

The exposure of honey bees to pesticides is an ever-changing problem for beekeepers, because each year new pesticides, as well as new formulations of the established ones, appear in the marketplace. The release of just one new chemical or different formulation has, at times, been devasting to honey bees. When Sevin (carbaryl) first was applied in orchards in the Northwestern United States, one beekeeper alone claimed to have lost several thousand colonies in less than a month. The heavy loss of colonies happened unexpectedly and so fast that a huge number of colonies were killed before remedial steps, such as moving the colonies, could be taken.

More recently in the same region, the change from the customary spray-form of methyl parathion to the new encapsulated form was blamed for the loss of several thousand colonies. Beekeepers were well aware of the highly toxic nature of methyl parathion, but they were not aware, or prepared, for the increased toxicity due to the greatly extended period over which the encapsulated chemical will kill bees.

Unfortunately, much of the information that beekeepers acquire on pesticides and honey bee mortality comes through personal observations when colonies are weakened or killed by new chemicals.

A tragic example of honey bee mortality was reported in Arizona, where the number of colonies dropped from 110,000 in 1964 to 53,000 in 1971—primarily because of the intensive, widespread cotton-spray program. Cotton has blossoms that are an attractive source of nectar over most of the summer. To protect the cotton, the farmer makes as many as a dozen applications of toxic insecticides to the plants during the summer. With few exceptions, bees cannot survive in this type of an environment (Moffet and others *1977*). Another

well-documented series of heavy bee losses due to pesticide poisoning comes from California, where beekeepers lost an average of 62,500 colonies *a year* from 1962 to 1973 (Atkins *1975*).

Since the manufacture of pesticides develops millions of dollars annually for agricultural businesses through domestic and foreign sales, and because crop farmers see no easy alternatives to pesticide application at the present time, what hope is there for the beekeepers' bees? Fortunately, there are methods of application, types of formulation, apicultural management practices, legislative measures, and other protective devices which can and do aid the honey bee and the beekeeper not only to survive but to pollinate crops and produce honey successfully in an environment often containing many poisonous chemicals.

The following information is presented to help the beekeeper better understand pesticides and to successfully meet the challenge of pesticides killing honey bees.

Classes of Pesticides [3]

The need of human beings to effectively control their environment is most evident in their agricultural pursuits. Modern farming covers large tracts of land under uniform planting, and this has made pest control mandatory. The evolution of pest control agents originated with natural products such as arsenicals, petroleum oils, and toxins derived from plants (nicotine and rotenone, for example). The advent of DDT, which was synthesized in a laboratory, heralded an era in which a mature chemical industry would screen synthetic chemicals for pesticidal activity. This effort spawned an impressive array of insect control agents. The selection of control chemicals is large. However, these materials can be grouped conveniently according to general chemical properties and modes of action.

Chlorinated Hydrocarbons

These include such important insecticides as DDT, BHC, toxaphene, and chlordane. The chemicals in this group are slowly reactive chemically, thus persistent in the environment. Biological degradation tends to be slow; hence, storage in fatty and muscle tissue causes these materials to become concentrated and enter our food chain.

---

[3] Toxicity of Pesticides to Honey Bees (Atkins et al. *1975*) lists many commonly used pesticides according to relative toxicity.

The mode of action of chlorinated hydrocarbons is still a subject of active research. They are classified as neuroactive agents which block the transmission of nerve impulses. Specifically, for example, DDT prevents the normal sodium-potassium exchange in the sheath of the nerve fiber— this exchange being the means by which a message is transmitted along the nerve. Because chemicals such as DDT are not very chemically reactive, it is felt that the mechanism of reaction with the sheath is not chemical, but rather that the size and shape of a DDT molecule may fortuitously permit it to fit into the proteins of the sheath. Such conceptualizations of the toxicological processes have promoted the search for new chemicals with better toxicological and environmental properties.

Organophosphorus Insecticides

These, today, account for about 30 percent of the registered synthetic insecticides/acaricides in the United States. They possess the common characteristic of inhibiting the enzyme cholinesterase, which mediates the transmission of nerve signals. Hence, organophosphates also are neuroactive agents. As their name implies, these materials contain phosphorus, and as a group they include parathion, Systox, DDVP, and malathion. They are quite reactive chemically and are not regarded as persistent in our environment, unless they are microencapsulated.

Carbamate Insecticides

These also are inhibitors of cholinesterase and feature a nitrogen-containing unit known to chemists as a carbamate function. Members of this class of insecticide include carbaryl (Sevin), baygon, Furadan, landrin, and zectran. For the most part, these materials are easily biodegraded and do not constitute the residual hazard of the chlorinated hydrocarbon class of insecticides. Interestingly, cholinesterase inhibition tends to be reversible for mammals and insects alike. A sublethal dose can bring on the usual symptoms of nerve poisoning (tremors, loss of muscular control, incontinence, vomiting), but the poisoned animal will return to normalcy in a very short time.

Other Pesticides

A wide variety of other synthetic chemicals may be applied to crops on which bees may be foraging. Herbicides and fungicides have bases for their activity which render them relatively much less toxic to honey bees. Still such materials are present in the biosphere of the honey bee, and little information is currently available dealing with the effects of these chemicals in combination with insecticides—a situation which occurs often under normal field conditions. Moreover, such materials as herbicides and nonconventional insecticides (such as insect sex attractants and insect growth regulators) to which bees are being increasingly exposed likely will be transferred to honey and stored pollen with, as yet, incompletely documented results.

The toxicity of a specific pesticide is a composite of its physical and chemical properties, the method of formulation (description follows), and the inherent ability of the honey bee to deal with the material internally. If the pesticide is of high volatility (an example is the fumigant TEPP), then the chemical may be absorbed through the bee's spiracles or respiratory system. Fortunately, TEPP is very quickly hydrolyzed—it reacts with moisture readily and is very nonpersistent. Absorption through the bee's integument is the basis for contact toxicity. The physical properties of an insecticide and especially of its formulation would be largely responsible for the relative hazard from this mode of entry into the bee. Ingestion of contaminated pollen and nectar offers yet another route of entry. The alimentary tract may become altered or paralyzed, making feeding impossible, or the bee's gut may cease to function. The ability of an insecticide to contaminate nectar and pollen would again be a composite of the physical/chemical properties of the material, its formulation, and the time of application of the spray relative to bloom.

Due to the extensive measurements of E. L. Atkins, C. Johansen, and others, a large amount of toxicity data has been collected, which allows an assignment of the inherent relative toxicity of the many pesticides now in use. These data are the results of both lab and field tests in which the materials usually were examined in their most common formulations. The organophosphates and carbamates are the most toxic to honey bees, with Furadan (LD_{50} 0.160μ g/bee) and parathion (LD_{50} $0.175\mu g$/bee) high on the list.

A number of chemicals were classed as only moderately toxic (LD_{50} 2–$11\mu g$/bee). Endrin, DDT, mirex, chlordane, Systox, and Phosvel are included in this group. It is of interest to note that several of these are chlorinated hydrocarbons. The lower toxicity of several of these chlorinated hydrocarbons may be due to an enhanced ability of the bee to degrade the compounds to nontoxic materials within its body. Resistance, or relative resistance, among insects often is an increased ability of the insect's biochemical constitution to chemically dispose of the insecticide.

Among the relatively nontoxic pesticides are allethrin, Kepone, Kelthane, and most of the fungicides and herbicides. Extensive studies have not been conducted with such control chemicals as pheromones and growth regulators. However, a great deal of toxicity data regarding their effects on mammals, birds, and insects other than bees has been collected which indicate nontoxicity. The growth regulators, on the other hand, may well affect the nature and volume of the brood. More studies involving these chemicals seem desirable.

Nonchemical Control

Along with the beneficial aspects of chemical pesticides come problems such as contamination of the environment and killing of beneficial insects, many of which are honey bees. To reduce agriculture's dependency on pesticides, a new concept of pest control has been developed called *integrated pest management* (IPM). Under IPM, all techniques and methods that are useful in controlling pests are used, including pesticides. However, a farmer applies a chemical pesticide only as a last resort. Primary reliance is on nonchemical controls, such as insect attractants, repellents, traps, insect-resistant plants, insect pathogens (disease), insect predators and parasites, time of planting, cultivation, time of harvest, sterilized insects, quarantines, and other practices. Not all of these techniques are utilized at the same time in controlling a pest; however, all of the control methods are considered noninjurious to honey bees, except for chemical control. Where possible, beekeepers should encourage farmers to use IPM techniques.

Application Methods

Once a compound is shown to be toxic to insects in laboratory tests, we must learn how to apply it efficiently to pests under field conditions. Considerable variations in toxicity toward a particular insect are sometimes seen when the insecticide is applied in different carriers. Therefore, the preparation of suitable formulations for an insecticide is a vital part of its development for practical use

and generally will determine the particular pest control situation in which it may be employed—as well as the degree of the danger to foraging pollinators such as honey bees.

At one time, there were more than 1,200 formulations in the United States based on DDT alone, and another 1,500 based on other chlorinated hydrocarbons.

Dusts can be made by mixing insecticides, which are solids (DDT, carbaryl), with an inert solid for a vehicle which sticks to the foliage. Because the exterior of bees is largely waxy and hairy, the dust adheres quite tenaciously to the bees as well. The insecticide may be dissolved in an organic solvent (xylene, kerosene) and the resulting solution sprayed. The solvent provides penetrating power, and the rate of evaporation of the insecticide from aerosolized portions of the spray may offer greater respiratory hazard to bees. An aqueous (water) suspension of the insecticide may be made if it is formulated as a wettable powder. Once the spray contacts the foliage, the water evaporates and exposes the insecticide to the environment in much the same way as a dust. Another formulation bears the name emulsifiable concentrate (EC). A solution of the insecticide in oil containing a detergent substance will form an emulsion (a suspension of finely divided liquid particles). Again, when the water dries from such a spray, the insecticide becomes exposed on the surface of the foliage. Such a spray has less penetrating power and is more evenly disseminated than the wettable powder formulation.

Microencapsulated Insecticides

Of considerable concern is the encapsulated insecticide formulation, such as Penncap-M, containing methyl parathion. The insecticide is dissolved in an organic chemical, then treated with another chemical with which it reacts to form a polymer. The insecticide molecules become imbedded in the polymer matrix, and the resulting free-flowing powder (having a particle size of 10 to 50μ) is sprayed as a water emulsion. After the water has dried, the particles behave much like a dust spray, with the exception that contact toxicity is rather minimal because the insecticide is within the particle (capsule). Toxic action is due primarily to release of the insecticide through the capsule walls as a vapor (gas). Beekeepers should be warned that bees will, in fact, carry the capsules to the hive along with pollen. Moreover, in field tests, foliage sprayed with Penncap-M remained toxic for a much longer period of time than did foliage sprayed with methyl parathion formulated as an emulsifiable oil.

While debate continues on the subject of the relative merits of different formulations for insect control and bee welfare, it is accurate to say that there is no safe formulation where honey bees are concerned, and each formulation has its unique hazards. A list of insecticide formulations in order of decreasing hazard to bees follows: Dust; wettable powder; flowable, emulsifiable concentrate or soluble powder or liquid solution; and granular formulation. The position of encapsulated insecticides has not yet been well defined, but may be more toxic to bees than are dust formulations. Because of the rather insidious nature of encapsulated formulations such as Penncap-M, however, beekeepers should be particularly cautious about exposing their foraging bees to such materials.

Pesticide Drift

Small particles of pesticides often become suspended in the atmosphere as a result of wind currents or heated air rising. This contamination of the air frequently kills honey bees, especially when the poison settles on plants where the blossoms are attracting bees. Frequently, the farmer applies the pesticide to a crop not attractive to bees, but the wind blows the poison onto a cultivated crop or to nearby weeds where bees are actively foraging for nectar or pollen, or both. The outcome can be catastrophic with many adult field bees dying (fig. 1). If the pesticide drifts or is blown into the entrance of a hive, many or all house bees and brood may succumb. Pesticide drift that is most damaging to bees usually originates in a field a short distance (1 or 2 miles or less) from the point of contact with the bee. Beekeepers, however, occasionally report drift over several miles (less than 10) that results in serious adult bee mortality. Long-range drift does occur,

FIGURE 1.—A double swirl of pesticide containing dust may effectively contain insect pests, but it also can be devastating to honey bee and wild bee populations—if the target plants are in bloom or the pesticide drifts onto a nearby crop that is being visited by bees.

but noticeable damage to adult bee populations is doubtful because of the enormous dilution factor.

Signs of Bee Mortality

Outside the Colony

The most obvious indication of heavy exposure to poisons is the heavy accumulation of dead or dying bees at the hive entrance and on the ground between the colonies (fig. 2). (In strong colonies, natural mortality of up to 100 dead adult bees per day is a normal die-off rate. When the rate exceeds 100 per day, then poisoning may be suspected.) Individual bees that have been poisoned, frequently are seen crawling on the ground near the entrance or twirling on their side in a tight circle. Others appear to be weak or paralyzed. These gross symptoms of poisoning vary with the type of pesticide and the degree of exposure. Foraging bees also may die in the field or on the flight back to the hive.

FIGURE 2.—Large-scale kills: Some pesticides kill bees in the field, while others kill the bees in or near the hive.

In severe cases, beekeepers and scientists have reported dead bees dropping like rain after an aerial spray of a contact poison, such as methyl parathion, had been applied to a field of blossoms during the middle of the day. Frequently, the bees that die away from the hive are difficult to find since they drop into vegetation or dry up and the wind blows them away. The inability to find the dead bees often leads to the false conclusion that the colonies did not suffer a loss of population. The useful aspect of bees dying away from the hive is that poisoned nectar and/or pollen are not brought into the hive, where they might be fed to the immature bees (brood) or stored in the combs. This may inadvertently protect the honey harvested for human consumption from pesticide contamination.

Inside the Colony

Bees die from various causes, and sometimes the cause of death may be difficult to determine. However, in moderate to heavy losses resulting from pesticide exposure, the problem often can be detected as soon as the hive is opened—the population is gone! Usually this loss occurs during warm weather when crops or weeds are being treated. Most frequently, the field bees are first to encounter the poison, but in more severe poisoning the house bees also die. When the house bees die, the brood will show signs of neglect or poisoning and many, or all, immature bees still in the cells may die. If a strong colony (three or more deep hive bodies) loses its foraging bees, nectar and pollen collection will be drastically reduced, but the population could recover in a few weeks. If the foragers and hive bees are lost, however, the colony may never recover and frequently will perish during the winter. The economic ramifications of a population loss depend on the season when the loss occurs. For example, if the field force is lost just before the colony is to be rented for pollination, or package bees shook, or the main nectar flow—the economic value of the entire colony would be lost.

Atypical Losses

Most pesticides are applied when the weather is warm and sunny. When a colony suffers a heavy loss of bees during favorable weather, pesticides frequently are suspected and usually are the cause of death. (Exceptions do occur, such as treatment of apple and almond blossoms during less favorable weather.) When conditions are cold and rainy, large numbers of bees often die, but not from pesticide exposure. More often, starvation, nosema disease, or disappearing disease account for the losses. Adult populations also may dwindle because of a poor queen or swarming. Sometimes, pesticides are blamed for losses due to these other factors.

Protection of Bees

Although the applicator of pesticides frequently is responsible for the poisoning of honey bees, the beekeeper should be aware of management techniques that can be used to lessen the damage or even, in some situations, avoid the problem. Several management practices are known that may help protect bees from pesticide-caused mortality.

Timing

Timing the application of a pesticide, and especially insecticides, can be extremely important. Treatments can and should be applied only when bees are not foraging for nectar or pollen. If bee-attractive plants need to be treated while in bloom, they should be treated at night or in the early morning or late evening when the bees are not flying. Recently, night treatments have been used successfully in southern California. Another new concept is to determine the time of day when each species of plant secretes nectar and treat the plants when they are not attracting bees (fig. 3).

Occasionally, the time of year is important—namely, treating when optimum control of a crop pest can be achieved and be least harmful to honey bees. As an example, alfalfa should be sprayed for weevil control in autumn rather than spring or not until after the first cutting in early summer. Finally, advise everyone to use pesticides only when absolutely necessary.

Relocation

Relocation of colonies may be desirable if a regime of repeated applications of insecticide is followed near an apiary. Information on the type of pesticides used and the frequency of application should be considered before moving colonies to a new location. Sometimes moving colonies, even for short distances or for brief periods 24 to 48 hours to avoid short-residual poison sprays and dusts, may be the best solution to the bee-mortality problem. Moving colonies, however, can be costly, not only in terms of vehicle expenses and labor, but often queens are killed or injured during the move and the "morale" of the colony may be disrupted for

FIGURE 3.—Many environmental factors contribute to a successful honey crop, but freedom from pesticide damage is a major one. When plants are treated with a pesticide while in bloom, the nectar and pollen may become contaminated. However, pesticides have not been found to contaminate honey stored in hives because foraging bees carrying poisoned nectar die before depositing the honey in their hive.

several days. Nectar storage can be reduced, apparently because of the need for the bees to reorientate to the new bee pastures. Unfortunately, some areas are dangerous enough to bees for a beekeeper to avoid altogether, thus requiring a permanent move to an area with lower pesticide exposure. (New areas with low or no pesticide exposure and satisfactory honey production are difficult to find.) Pesticide-free sanctuaries for bees have been tried in the Northwestern United States, but they have proved impractical or unsuccessful.

Field Protection Techniques

Field protection of bees to alleviate damage by insecticides can be accomplished in several ways.

One successful technique is to cover each colony with a large burlap sack shortly before the insecticide is applied. Covering the hive entrance with the burlap prevents foraging during the day. The cover is removed at the end of the day when it is too dark for the bees to forage. Where colonies are covered, water should be applied to the burlap every 1 to 3 hours to cool the exterior portions and as a water source for the bees cooling the interior of the hive. A new device, the Wardecker Waterer (Moffett et al. *1977*), which provides a reservoir of water within the hive, proved successful in supplying water to covered colonies in the Southwestern United States during periods when air temperatures were 100°F or higher for several weeks. This waterer is basically a modified shallow super. Colonies equipped with this device need no water applied to the burlap cover during confinement.

Whenever possible, it is always best to locate beeyards outside the spray/dust pattern rather than within the treated field. Moreover, colonies weakened by insecticide treatment should be fed sugar syrup and pollen cake to stimulate brood production as an aid to population recovery. Frequently, weak colonies must be united to save the remaining bees and brood or a queenless package of bees added to the damaged colony to strengthen the population.

If a pesticide carried by the bees infiltrates the combs and contaminates the nectar or pollen, or both, then the combs may need to be replaced before a colony will recover. Contaminated combs either may be soaked in water to help in the removal of stored pollen or the entire comb melted and replaced with wax foundation. Before washing or melting combs, however, a sample of the pollen, wax, and honey should be analyzed chemically to determine the amount of pesticide residue present, if any.

Another form of protection is direct and effective communication between beekeepers and local farmers. When farmers are educated to understand the complexities of a bee business, they often are more sympathetic about the problems of beekeepers, such as insecticide damage. Beekeepers should make their presence known as well as the location of each beeyard. Register the location of each apiary with the State bee inspector or other designated officials. In addition, post a sign in each apiary with the name and address of the owner.

Resistance to Pesticides

Genetic stocks of honey bees that have a higher level of tolerance to certain pesticides have been developed. None of these stocks of bees, however, has been developed commercially—mainly because resistance to one chemical poison would not necessarily protect the bees from other pesticides, and any heavy exposure of "resistant" bees still would cause bee mortality. Incorporating many desirable genetic characteristics, such as gentleness, good honey production, nonswarming, successful wintering, and pesticide resistance, into a stock of honey bees is difficult and expensive.

Selective Toxicity of Pesticides

An important facet in protecting bees is to educate pesticide fieldmen and pesticide applicators concerning the problems beekeepers face from pesticides—and to encourage the applicators to treat a field at a time and in a way least injurious to honey bees. They should always use the least toxic chemical that will achieve the desired results. This education can be accomplished through personal contacts, magazine and newspaper articles, and possibly best of all by discussion sessions between applicators, beekeepers, pesticide fieldmen, and apiculture scientists organized by the county agricultural agent or someone in a similar position of agricultural leadership.

Repellents

Chemical repellents have been studied for many years, especially by E. L. Atkins of the University of California. The repellent is added to the pesticide before field application and is intended to discourage bees from visiting plants until the pesticide becomes relatively nontoxic. Field tests showed several compounds to have repellency, but more research is needed before they are used commercially by farmers.

Bee Recovery

When colonies sustain the loss of part or even most of their foraging bees, most frequently the beekeeper need only wait a couple of weeks and "new" bees in the hive will emerge and take over the field duties.

In severe mortality where all the field bees and many house bees have been lost, the beekeeper will need to resort to one or more of the following management techniques: Feed pollen and/or pollen substitute, add a queenless package, feed sugar syrup, move to an area with at least some natural nectar and pollen available that is relatively free of all pesticides (these areas are difficult to find nowadays), or unite two weak colonies. Sometimes, the contaminated combs must be removed and replaced with unexposed combs or frames of foundation. Weakened colonies should be protected from factors causing stress—such as cold temperature, excess heat, and lack of water. Factors favoring increased brood rearing aid colonies in their recovery.

Programs to Aid Beekeepers

Honey Bee Indemnity Program

For many years, beekeepers lost large numbers of colonies following the application of insecticides that had been approved by the Federal Government. Since the beekeepers received no compensation for these losses, the Government enacted national legislation to partially repay each beekeeper for pesticide-killed bees. Beekeepers who exercised reasonable precautions to avoid pesticide damage but still lost bees could apply for indemnity payments after January 1, 1967. The main goal of this program was to aid the bee industry in remaining financially stable and to ensure that enough strong colonies would be available to pollinate agricultural crops nationwide. This goal has been accomplished.

The Agricultural Stabilization and Conservation Service (ASCS) administers the indemnity program. When beekeepers have a loss, they should contact the nearest county ASCS office immediately. An ASCS inspector will check the colonies and determine the extent of the damage. The inspector also will assist the beekeeper in filing a claim. The payment is based on the population size in a colony before and after exposure to the pesticide. The indemnity program, however, is not a substitute for good bee management designed to avoid areas of heavy application of pesticides.

To accurately establish that a loss occurred, beekeepers should maintain detailed records of their colonies, noting by date such items as colony condition, population size, syrup and pollen feeding, and honey production. The more detailed the records, the easier it is to establish the true magnitude of a loss and receive reasonable compensation.

Pesticide Detection and Analysis

Residues of pesticides may be present as the original pesticide or as an identifiable degradation product, or both. Frequently, the amount of residue involved is extremely small. Analysis of these residues consists primarily of:

(1) Blending and extracting the biological material (such as bees and pollen) with a suitable solvent system so as to maximize the recovery of those pesticides whose presence is suspected, and their metabolities. This eliminates the bulk of the biological substrate.

(2) A series of liquid-liquid extractions and column chromatographies planned so as to further separate the residues from other materials of biological origin.

(3) Detection of the residues at the highest possible sensitivity to avoid interferences from substances not previously removed. A very fast method for cleanup, which is increasing in use, is gel permeation chromatography. Large molecules of biological origin emerge from the column first, thereby trimming the crude extract down directly to a much cleaner sample.

The most popular detection device is gas-liquid chromatography, wherein the residue-containing sample is volatilized and chromatographed as a vapor (fig. 4). A variety of detectors may be employed at the exit of the chromatography column, including element-specific types which will "see" only chemicals containing nitrogen, phosphorus, or sulfur. Since the presence in biological specimens of interfering substances containing these elements is unlikely, procedures employing specific detectors generally require less initial "cleanup" and are growing in popularity.

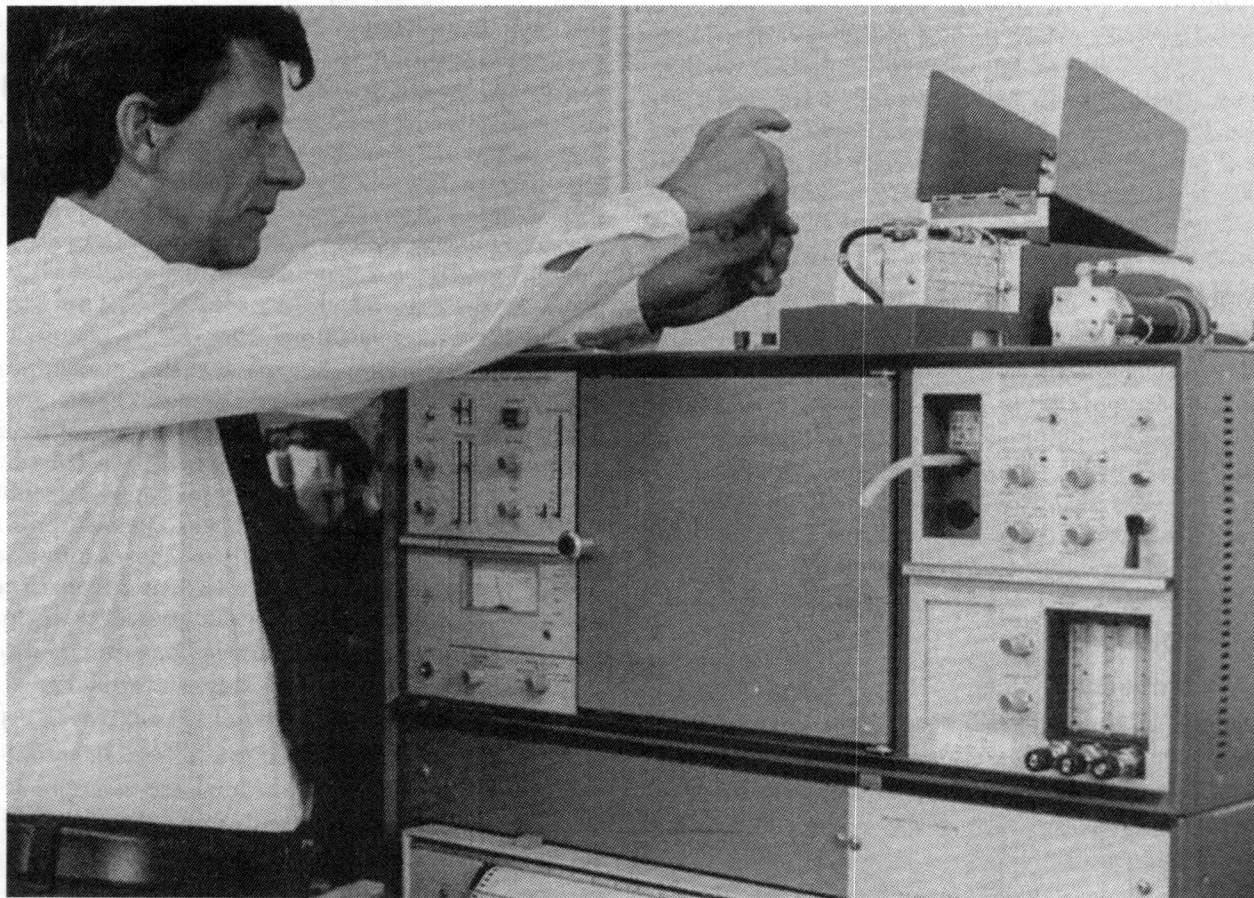

FIGURE 4.—A gas chromatograph is used by scientists to detect minute quantities of pesticide residue in bees and bee products and on plants.

Thin-layer and paper-chromatography also are useful for establishing the identity of pesticide residues. The ultimate visualization for detection usually involves spraying the chromatography plates with a chemical that reacts with the pesticide/metabolite to produce a characteristic color. Unfortunately, the method is less useful for quantitation.

Mass spectroscopy also is used for residue analysis. Currently, the method lacks the sensitivity required, and the cost of equipment prevents its more general use. However, fluorometric methods are gaining adherents because of improved sensitivity and the rather minimal sample preparation required. The residue must be capable of absorbing visible or ultraviolet light. The intensity of the reemitted light is measured at some suitable wave length. This intensity can be compared with standards and functions as a quantitative measure of the residue. Current research centers on altering nonfluorescing pesticides so as to render them fluorescent and hence detectable by this method, and on applying high-pressure liquid chromatography employing ultraviolet detecting equipment for rapid cleanup, detection, and quantitation.

Government Regulation

For many years, governments have legislated the amounts of poisonous chemicals permitted in foodstuffs for human consumption. In the United States and other countries, there are stringent regulations and requirements on pesticide manufacturers to produce toxicity and persistence data before they are allowed to market any pesticide. One difficulty of such legislation is the wide divergence of opinion among pharmacologists and toxicologists as to the hazard of any pesticide. This can be demonstrated by the variation in tolerances set on the organochlorine insecticides by different countries.

The Environmental Protection Agency (EPA) was created in 1970 in response to public concern over pollution of the environment by substances including pesticides. EPA is "to prevent and abate degradation of the environment and to promote environmental enhancement" (*Federal Register* 1971). This independent agency acquired the authority to monitor and license pesticides (formerly a USDA assignment). EPA thereby regulates the use of all pesticides in the United States through the registration of these materials. EPA also sets tolerance levels in food (formerly a Food and Drug Administration (FDA) assignment) and controls radiation standards (formerly an Atomic Energy Commission (AEC) assignment). FDA, however, retained the power to regulate the final limits of pesticide residues in foods. This is accomplished by spot-checking, sampling, and occasional confiscation. USDA has retained the power to regulate the pattern of use for agricultural pesticides through recommendations resulting from field testing by USDA and university scientists and from State extension specialists.

Pesticide Applicator Certification

In 1972, Congress passed an amended version of the Federal Insecticide, Fungicide, and Rodenticide Act (FIFRA) and gave the EPA the responsibility for carrying out the provisions of the act. Under the act, many people who apply pesticides are required to pass an examination or otherwise demonstrate their knowledge on proper use of pesticides. The Federal Government requested that each State set up a program to certify people such as commercial applicators (aerial and ground equipment) and even certain farmers who apply *restricted use* pesticides. (These pesticides are many organophosphates and some carbamates, depending on dosage, formulation, and site of application.) Many States are conducting certification courses or workshops to update applicators' knowledge of pesticides and to aid them in qualifying for certification. Certification under this act should benefit the beekeeping industry through more judicious use of pesticides.

Research on Pesticides

The chemical-manufacturing industry frequently markets new pesticides or new formulations of previously marketed compounds. These new products often create new or additional problems for beekeepers by killing bees. Consequently, research programs have been established in universities and at SEA bee-research laboratories to find practical ways of reducing or avoiding honey bee mortality. One recently discovered method for reducing mortality is the Wardecker waterer device, which provides bees with an internal source of clean and readily available water for heat regulation of hives and nutrition. Another method that is currently being researched is use of chemical repellents mixed with pesticides that deter bees from entering fields.

State Laws Protecting Bees

Some States have laws to help protect honey bees from pesticide exposure. Such laws encourage the use of the recommended amounts of pesticide, application of pesticide when plants are not highly attractive to bees, abstention from application when climatic temperatures are high and wind velocity above 5 to 10 mph, aerial application in the early morning or late evening only, and notification of beekeepers 1 day or more before pesticide application near their colonies. Beekeepers in some States are required to *register the location of their beeyards,* so that a notification program can be utilized.

Sources of Information on Honey Bee Exposure to Pesticides

Often one of the following individuals or groups can answer questions or render aid:
— County agricultural agent
— State bee inspector
— State entomologist
— State department of agriculture
— Department of entomology, State university
— Honey bee laboratories, USDA
— State and national beekeeping organizations
— Pesticide manufacturing companies
— County Agricultural Stabilization and Conservation Service of USDA

References[4]

ANDERSON, L. D., and E. L. ATKINS.
1968. PESTICIDE USAGE IN RELATION TO BEEKEEPING. Annual Review of Entomology 13:213–238.
ATKINS, E. L.
1975. INJURY TO HONEY BEES BY POISONING. *In* The hive and the honey bee. p. 663–696. Dadant & Sons, Hamilton, Ill.

---

[4] These publications usually are available from college and public libraries.

———L. D. ANDERSON, and E. A. GREYWOOD.
1970. RESEARCH ON THE EFFECT OF PESTICIDES ON HONEY BEES 1968–69. Parts I and II. American Bee Journal 110:387–389; 426–429.
———L. D. ANDERSON, H. NAKAKIHARA, and E. A. GREYWOOD.
1975. TOXICITY OF PESTICIDES TO HONEY BEES. University of California, Division of Agricultural Science, Leaflet 2286.
———L. D. ANDERSON, D. KELLUM, and K. W. NEUMAN.
1976. PROTECTING HONEY BEES FROM PESTICIDES. University of California, Division of Agricultural Science, Leaflet 2883.
BROOKS, G. I.
1974. CHLORINATED INSECTICIDES, VOL. I, TECHNOLOGY AND APPLICATION. Ch. 2, p. 18–25. CRC Press, Inc.
JOHANSEN, C. A.
1966. DIGEST ON BEE POISONING, ITS EFFECTS AND PREVENTION. Bee World 47:9–25.

———

1977. PESTICIDES AND POLLINATORS. Annual Review of Entomology 22:177–192.
KLEINSCHMIDT, M. G.
1972. INSECTICIDE FORMULATIONS AND THEIR TOXICITY TO HONEY BEES. Journal of Apicultural Research 11:59–62.
MATSUMURA, F.
1975. TOXICOLOGY OF INSECTICIDES. Chapters 3, p. 47–103; 4, p. 105–163. Plenum Press, New York.
MOFFETT, J. O., A. STONER, and A. L. WARDECKER.
1977. THE WARDECKER WATERER. American Bee Journal 117(6): 364–365; 378.
STONER, A. and A. L. WARDECKER.
1978. REDUCING INSECTICIDE LOSSES TO HONEY BEES FROM COTTON SPRAYING. Journal of Economic Entomology.
NATIONAL ACADEMY OF SCIENCES.
1975. PEST CONTROL· AN ASSESSMENT OF PRESENT AND ALTERNATIVE TECHNOLOGIES. 506 p. vol. 1, Report of the Executive Committee, National Academy of Sciences, Washington, D.C.
TODD, F. E. and S. E. McGREGOR.
1952. INSECTICIDES AND BEES. In U.S. Department OF Agriculture, Yearbook of Agriculture (Insects) 1952:131–135.

HYMENOPTEROUS INSECT STINGS

By Thor Lehnert [1]

The honey bee is the most common single species of stinging insect in the United States. Bees and the related wasps, hornets, yellow jackets, and ants usually do not sting unless stepped on, touched, or molested. They are usually not active at temperatures below 55° F or on rainy days. The highest incidence of stings is in August. Yellow jackets, honey bees, and wasps are the worst offenders, in that order. Yellow jackets cause more moderate and also more severe reactions than bees or wasps.

Bee Venom

The poison gland system of the bee consists of a small alkaline gland and a larger acid gland. The venom comes from these two glands. The stinger is made up of two lancets with sharp barbs pointing backward, similar to a harpoon. When a honey bee stings, its stinger is pulled from its body. Even after the stinger is separated from the bee, its muscular mechanism can continue to force its venom into the wound. Because venom continues to be injected by the stinging mechanism, the stinger should be removed immediately. Most people make the mistake of pulling it out with the thumb and index finger and thereby squeeze more venom into the wound. The stinger should be scraped or scratched out with a fingernail.

Other stinging Hymenoptera, such as yellow jackets, wasps, and hornets, retain their stingers.

Bee venom is a water-clear liquid with a sharp, bitter taste and a distinct acid reaction. The specific gravity is 1.1313. The venom is easily soluble in water and acid, but almost insoluble in alcohol.

The three toxic effects of bee venom are neurotoxic (paralysis of the nervous system), hemorrhagic (increase in the permeability of the blood capillaries), and hemolytic (destruction of red blood cells).

[1] Research entomologist, Science and Education Administration, Bioenvironmental Bee Laboratory, Beltsville, Md. 20705.

Recent work reveals that bee venom is a very complicated substance with several active biochemical components. At least eight active components plus several biological inactive components have been identified. The substances showing activity are histamine, melittin (a protein), a hyaluronidase, and phospholipase A.

The histamine recovered was shown not to be a major pharmacological factor in bee venom. However, histamine also is released from body cells because of the reaction to the sting in allergic persons.

Melittin, a protein having a molecular weight of 33,000 to 35,000, is thought to be responsible for the general local toxicity of the venom. Melittin in high concentrations also has caused hemolysis of red blood cells.

Bee venom contains at least two enzymes—a hyaluronidase and phospholipase A. The hyaluronidase is believed to be the "spreading" factor. By breaking down the cell-cementing substance, hyaluronidase allows the toxic principles of bee venom to infiltrate the tissues.

Phospholipase A apparently has no general toxicity. However, through indirect action on the unsaturated fatty acids, it causes hemolysis of red blood cells. Phospholipase A also causes inactivation of thrombokinase, inhibits oxidative phosphorylation, and attacks enzymes involved with metabolic dehydrogenation. The pain experienced after being stung may well be the result of these last three actions.

For many years, formic acid erroneously was believed to be the major component of venom produced by the honey bee, and this belief is still held by many. The action of venom is much more complex than the simple concept of direct action on the tissue by formic acid.

Sting Reactions

Usually insect stings cause only a local reaction, with pain lasting for several minutes after penetration of the stinger. A redness and slight swelling at the sting site also may occur. Until recently, some people believed severe symptoms and death from stinging insects were due to venom being introduced directly into a small blood vessel. Severe reactions are now considered to be due to sensitivity to proteins in the venom.

In about 2 percent of persons, a hypersensitivity develops in which each additional sting produces a more severe reaction. Hypersensitivity may appear after a varying number of stings, usually each sting making the reaction progressively worse. Some develop sensitivity after one sting, whereas others after a series of normal reactions. In a few individuals, hypersensitivity appears to be inborn. The first sting has resulted in death.

Symptoms in an allergic person usually appear within a few minutes after the sting, but may not appear for 24 hours. Local swelling may be excessive. A hivelike condition may break out over the body. There is a sensation of choking, difficult breathing, asthma, and the lips turn blue. Shocklike symptoms, vomiting, and loss of consciousness may follow in rapid succession.

Treatment and Precautions

Treatment by a physician is divided into three stages: (1) Immediate treatment for anaphylaxis is epinephrine 1:1,000, 0.3 cc to 0.5 cc, given intramuscularly or by deep subcutaneous injection. (2) Second-stage treatment includes a sympathomimetic agent, such as metaraminol, 100 mg in 500-cc isotonic saline solution given intravenously, antihistamine given intramuscularly, and corticosteroids. (3) Long-term care involves immunization with the appropriate insect antigen.

The long-term treatment undertaken to relieve the allergic condition to stings is known as desensitization. This treatment consists of a graduated series of injections of an extract made from the body of the offending type of insect. If the stinging insect is not identified, the sensitive person should be treated with an antigen composed of honey bee, hornet, paper wasp, and yellow jacket. Many allergists are now getting better desensitizing results with pure venom.

The degree of relief from allergic reactions is not the same for all persons after desensitizing treatments. The frequency of treatment also varies. For some people, one series of treatments is effective. For others, the treatment must be repeated with booster injections once a month.

Another important point is that of crossprotection. The injection of stinging insect extract of one species will protect some persons against all stinging insects. Some will need a combination of extracts to achieve complete protection. This makes the procedure of desensitization an individual process.

Desensitization seems to be helpful to about three-fourths of persons with severe reactions, in that they report lessened reactions to subsequent stings. Desensitization should be considered by persons who have severe reactions.

Such persons should have an emergency kit available at all times. This kit should contain ampules of epinephrine 1:1,000 and an injectable antihistamine. Some doctors now recommend the use of aerosol bronchodilator sprays which contain adrenaline-like compounds for persons sensitive to stings. However, the person who is extremely allergic and becomes faint and lapses into unconsciousness a few minutes after a sting must be taken to a physician or hospital to receive immediate treatment. Instructions on the use of first aid remedies should be obtained from a physician.

A sensitive person should be immunized, since there is no way of completely avoiding stinging insects. However, one can take several precautions to prevent stings. For example, the type of clothing worn affects the probability of being stung. Beekeepers consider clothing color to be one of the most important factors. Light-colored clothing should be worn in preference to dark, rough, or woolly clothing. Bees are attracted to suede or leather materials, particularly horsehide. If around bees, women should wear some type of head covering to keep the bees out of their hair.

The odor of perspiration has little effect on bees. Certain hair oils and perfumes should be avoided because they attract bees, which may sting after being attracted.

Bees will react more quickly to a moving object than to an immobile object. This can be demonstrated easily by making fast, jerky hand movements when working with a hive of bees. Striking, swatting, and swinging at bees will increase the chance of being stung.

No success has been attained in developing an effective repellant against stinging insects. Dimethyl phthalate, which has been effective in repelling mosquitoes and biting flies, is not effective as a repellant against stinging insects.

Only a small percentage of the population suffers serious reactions from hymenopterous insect stings. However, when reactions other than local irritation and swelling do occur, a physician should be called for immediate treatment. He should be consulted for long-range protection with desensitizing shots.

Yellow jacket wasps sting readily and often are confused with honey bees. The wasps build nests in the ground and will sting when people walk or stand near the entrance, which may be difficult to locate. Stings attributed to honey bees are often caused by yellow jackets.

References

AMERICAN MEDICAL ASSOCIATION.
    1965. INSECT-STING ALLERGY. American Medical Association Journal 193:115–120.
FRANKLAND, A. W.
    1976. BEE STING ALLERGY. Bee World 57:145–150.

HODGSON, N. B.
    1955. BEE VENOM: ITS COMPONENTS AND THEIR PROPERTIES. Bee World 36 217–222.
MARSHALL, T. K.
    1957. WASP AND BEE STINGS. Practitioner [London] 178 712–722.
MORSE, R. A., and R. L. GHENT.
    1959. PROTECTIVE MEASURES AGAINST STINGING INSECTS. N.Y. State Journal of Medicine 59:1546–1548.
PECK, G. A.
    1963. INSECT STING ANAPHYLAXIS. California Medicine 99:166–172.
SCHOFIELD, F. W.
    1957. PREVENTION OF SEVERE REACTIONS FOLLOWING STINGS OF THE HONEY BEE. Canadian Bee Journal 65:11–13.
SHAFFER, J. H.
    1964. HYPOSENSITIZATION WITH INSECT EXTRACTS. American Medical Association Journal 187:968.
STIER, R. A. and R. F. STIER.
    1959. THE RESULTS OF DESENSITIZATION IN ALLERGY TO INSECT STINGS. American Academy of General Practitioners 19:103–108.

MANAGEMENT OF WILD BEES

By Frank D. Parker and Philip F. Torchio [1]

The term "wild bee" is used commonly for all bees except honey bees in the genus *Apis* (hence, apiculture or culturing of honey bees). Many persons also refer to feral honey bees as wild bees, so the term is somewhat ambiguous. Bees generally are distinguished from other flying hymenopterous insects by their characteristic plumose body hairs. Bees are of many sizes, shapes, and colors. Some of the smallest bees, *Perdita*, are less than 3 mm, whereas our largest leafcutter bee is over 80 mm. Almost the entire range of colors is found among the brightly marked bees, including many beautiful metallic species.

Almost anywhere in the United States, one can easily observe many species of bees actively visiting flowers for nectar and pollen or engaged in the processes of constructing nests. There are approximately 5,000 species known to occur in North America, plus an estimated 1,500 species not yet described. Wild bee experts agree there are at least 30,000 species of bees in the world. This number of species is more than all the fish, bird, and reptile species combined.

Most bee species construct either single or complex nests in the ground. Some make earthen, leaf, or resin nests on rocks and plants. Other bees make or utilize crevices in rocks or plant stems, insect borings, and plant galls for their nesting sites.

Most bees live a solitary existence—each female after mating locates and builds her nest without the aid of other bees, and usually at a distance from her sister bees. However, some bees are quite gregarious and nest close to one another, sometimes in dense populations of up to a million nests in a few acres of soil.

Some bees prefer to nest at the same site year after year, but others relocate their nests each season. A small percentage of our wild bees are social or semisocial; that is, there is a division of labor among the bees occupying a single nest. Our total knowledge of the habits of bees in the United States is quite limited; less than 10 percent of their biologies have been observed and recorded.

Value of Wild Bees as Pollinators

One cannot easily place a dollar figure on the value of wild pollinators, simply because total impact on the environment is not known. The potential for utilizing, encouraging, or maintaining populations of wild bees has been barely researched. It has been calculated that the mere weight of wild pollinators outnumbers honey bee populations by hundreds of times. Studies on the impact of each pollinator species on fruit or seed production of our major crops is almost non-existent.

We know that transfer of pollen from one plant to another or from one plant part to another part of the same plant is essential for the reproduction of most flowering plants. Without pollinators, most of our native flowering plants would decline, disappear, or be replaced by nonflowering weedy species. Yet, reproduction each season of the multitude of wild flowering plants is often taken for granted to aid in maintaining soil moisture and fertility, and to provide food not only for wild life but for our domestic livestock as well. How many billions of dollars are these benefits worth?

It is easy to document the value of crop species visited by bees, but here again the importance of wild bees as crop pollinators has been sorely neglected. It has long been the general consensus that honey bees adequately pollinate crops and there is little need for wild bees. Unfortunately, such statements are premature since adequate research on the economic benefits of wild pollinators has not been done. Conversely, the research completed on the few wild pollinator species thus far studied has returned much compared with its investment costs.

---

[1] Research leader and research entomologist, respectively, Science and Education Administration, Pollinating Insect-Biology Management, Systematics-Research, Logan, Utah 84322.

The dependence on one species for crop pollination sometimes creates problems such as now being experienced in our almond industry. In California, there are not enough colonies of honey bees available to effectively pollinate the total almond acreage, and it is steadily increasing. Honey bees must be transported from as far as Minnesota and Canada, which substantially increases pollination costs and results in higher costs to the consumer. It seems wise to make greater efforts to study, conserve, and try to manage as many species of wild bees as possible.

There are several crops that are underpollinated by the honey bees, either because the bees are not physically adapted to pollinate them or the crops are not attractive to honey bees. Some of our most important crops, valued at billions of dollars, are in this category. These crops are alfalfa, soybeans, cotton, vegetable seed, and sunflowers, each of which is adapted to specific types of pollinators.

Recent research on the utilization of several species of wild bees as crop pollinators is just beginning to indicate some of their economic benefits. Currently, there is a viable multimillion dollar industry centered around the manufacture and sale of equipment, propagation, and pollination by contract of one of these wild pollinator species, the alfalfa leafcutter bee.

The alkali bee was the first wild bee to be utilized as a crop pollinator in the United States beginning in the early 1950's. Since that time, the alfalfa leafcutter bee and the blue orchard bee have been domesticated as crop pollinators.

Leafcutter Bees

There are many species of leafcutter bees that visit blooming alfalfa, but for the most part, our native species have not increased their populations to the point of being manageable. However, several Eurasian species have become established in the United States, and *Megachile rotundata* has become the principal alfalfa pollinator in several Western States. The alfalfa leafcutter bee, previously called *Megachile pacifica*, arrived on our east coast during the early 1930's and has spread to most parts of the United States and northern Canada.

In the 1950's, alfalfa seed growers noticed large populations of leafcutter bees pollinating their fields and began to increase the bees by providing nest holes in various kinds of wooden objects. By the 1960's, research at government facilities and western experiment stations had developed practical means of culturing large populations of the bees. Seed growers who managed these bees began to see significant increases in alfalfa seed yields. Leafcutter bees were particularly effective in areas that had no alkali bees. During the 1970's, intensive management practices have resulted in dependable leafcutter bee populations that ensure adequate pollination of western alfalfa seed fields.

The alfalfa leafcutter bee, about half the size of the honey bee, is black with white-yellowish bands on the abdomen (fig. 1). These bees are particularly fond of sweet clover and will visit it to the exclusion of alfalfa. Other crops visited for pollen are forage legumes, mints, crucifers and many weeds or garden plants. The adults commonly are found flying about near outbuildings, fence posts, cliff banks, or other suitable nesting sites. These leafcutter bees can utilize almost any small holes and will commonly plug small tubing, electrical sockets, and nail holes. Other favorite nesting sites are between the siding on frame homes and between or under the shakes or shingles on buildings. With such numerous nesting sites available, it is easy to understand why these bees became quite common and how large populations can be obtained for agricultural use.

Biological Features

Leafcutter bees are advantageous for alfalfa pollination because they:

(1) Usually forage within fields where they nest, making them less susceptible to being killed by pesticides applied to adjacent fields, and likely will not pollinate fields owned by other growers;

(2) Collect pollen from and trip the alfalfa flower readily at the rate of 8 to 10 florets per minute;

(3) Forage about the same time alfalfa blooms;

(4) Are predictable in incubation of adult stages;

(5) Have a long field life—up to 9 weeks—and a high rate of reproduction (maximum of 39 cells);

(6) Are gregarious and nest in manmade objects;

(7) Select older leaves for nesting and are not destructive of shrubs and trees;

(8) Have sturdy leaf cells and cocoons and thus are suited to mechanized management operations.

Populations

There are no consistent numbers of leafcutter bees used to pollinate alfalfa. The rate depends primarily on each seed grower's management

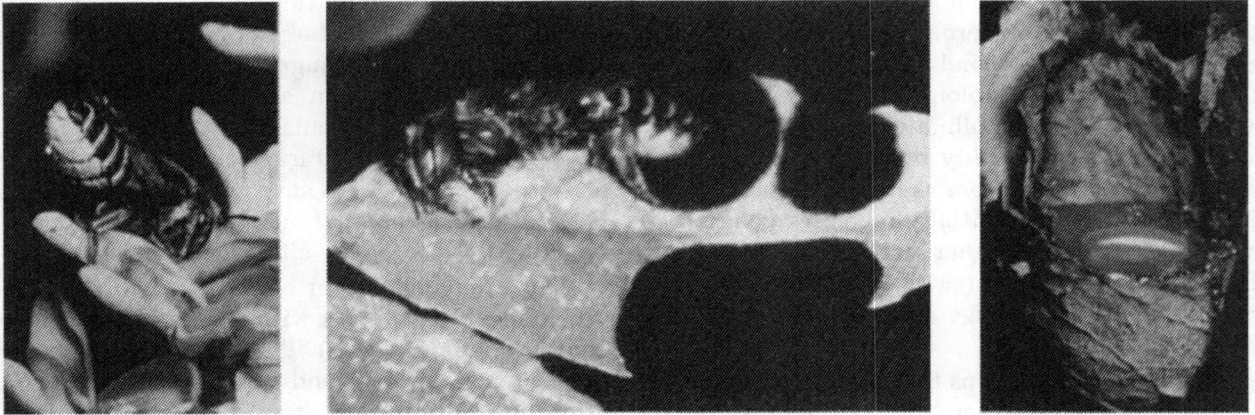

FIGURE 1.—Alfalfa leafcutter bee: *Left,* Female carrying pollen on the under side of the abdomen; *center,* cutting a leaf for nest material; oblong pieces are used for the sides of cells, circular pieces for capping the cells; *right,* egg floating on larval food supply of pollen and nectar at the base of the cell.

practices. However, it has been calculated that as few as 2,000 females per acre can adequately set up a field. The maximum carrying capacity has been estimated at 14,000 bees per acre. Early management recommendations were that small populations of bees be placed throughout the field for uniform coverage. This practice has changed gradually to the present concept of "mass pollination," whereby large populations of leafcutter bees (400,000) are placed on small acreages (25 to 50 acres) for a few days and then removed. The primary advantages of this type of management are that adequate pollination can be obtained in a short period, the bees can be used to pollinate additional fields, and fields can be treated regularly for pest insects without harm to the bees.

Nesting Materials and Shelters

Many kinds of nesting media have appeared during the last two decades of leafcutter bee management, and there are many claims of success. Substantial population increases have been obtained in several types of nesting media, but the most commonly used materials are boxes of soda straws, drilled boards (with or without removable backs), grooved boards, plastic wafers, and rolled-cardboard units. There are advantages and disadvantages to each type used. These characteristics are shown in table 1, although grower choice—unproved data from bee production—appears to determine which media are used.

Field shelters have changed from the initial small A-frame capable of holding a few boards to the large, self-contained, incubation-emergence type mounted on a trailer and capable of holding several hundred nesting units (fig. 2). It was assumed that facing the shelters to the east promoted increased bee activity, but research demonstrated that this practice actually increases mortality of young bee larvae. Sunlight penetrates the nesting holes and heats the larvae to as high as 130°F by 10 a.m. It is recommended that no sunlight actually strike the nesting units.

Some additional recommendations for shelters are:

(1) Use the conspicuous, larger sizes to attract and keep bees at the shelters;

(2) Ventilate each shelter to prevent an accumulation of heat at the top or through the sides;

(3) Place chicken wire or grills across the open end of the shelter to provide protection from birds;

(4) Remove debris from emerging bees, nest cleaning, and leaf drop from the floor to prevent an increase in scavenger beetles or moths;

(5) Mount shelters on a trailer with wide tires so movement does not jar the bees.

Storage During the Winter

Most growers store their bees as overwintering larvae in cold rooms set at 36° to 40°F. The leaf cells are removed from the nesting media (loose cells) or are left intact, depending on the kind of

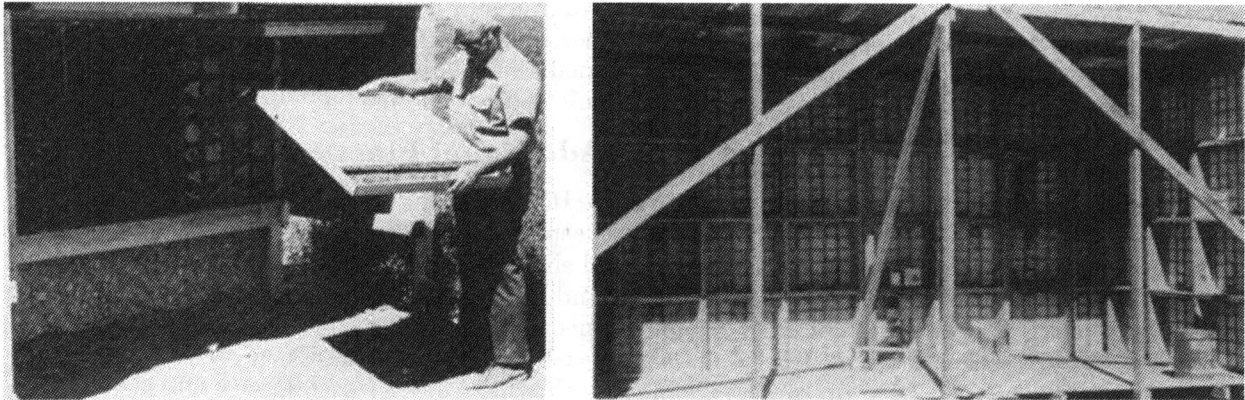

FIGURE 2.—Field shelters for alfalfa leafcutter bees: *Left,* The loose-cell technique of stocking shelters; *right,* a mobile field shelter that holds 400,000 nests.

TABLE 1.—*Relative desirability of various nesting materials for* Megachile rotundata [1]

Characteristic	Drilled boards	Plastic blocks with holes	Drilled boards, removable backs	Grooved wood	Grooved poly-styrene	Paper straws	Plastic straws	Corru-gated paper
Low cost	++	−	+	+−	+−	+++	++	+++
Availability	++	−	++	−	++	+	++	+
Strength (sturdiness)	+++	++	++	++	+−	−	+	−
Light weight	−	+	−	−	++	++	++	++
Compactness	+	+	+	+	+	++	++	++
Insulating property	++	++	++	+	++	+	+	+
Ventilation qualities	+	−	+	++	−	+	−	+
Separability from cells	−	+	++	++	++	−	−	−
Cleanability for reuse	−	+	++	++	++	−	−	−
Inspectability of contents	−	+	++	++	++	+	+	−
Resistance to cha cids	++	+	++	[2] −	+−	−	+	−
Attractiveness to bees	++	+−	+	++	+−	++	+−	+−
Safety from most birds	++	+	++	++	+	+−	+−	(?)
Safety from mammals	++	−	++	++	+−	−	+	(?)
Ease of storing [3]	+	+	++	++	++	+	+	+

[1] +++ (Excellent), ++ (Good), + (Fair), − (Poor), +− (Mixed reports).

[2] Assuming some warping of boards and poor fit of backing.

[3] Principal factor is storage separately from cells; secondary factors are compactness. Assumption made that holes are well occupied.

management used. Cold treatment prevents a buildup of scavenger beetles, moths, and other nest destroyers that can damage unprotected nests. Although this practice is widely used, caution is needed so bee larvae are not stored too early in the season (before they spin a cocoon and change to the overwintering stage). To prevent injuring bees by early storage, the nesting media should not be placed in cold storage for at least 2 weeks after removal from the field.

Generally, if nests are stored during the summer, larvae destined for the second generation will die, causing considerable problems in the emergence pattern the next season. Also, bee larvae stored at constant low temperatures for long periods have a higher death rate (10 percent) than those left at outside temperatures (in Utah). So growers should determine the abundance of nest destroyers and their potential for destruction before long-term cold storage of larvae.

Incubation

Growers must decide when bees will be needed in the spring to begin incubation of larvae at temperatures averaging 86° to 90°F approximately 20 days before bee emergence. The most critical factor during the incubation process is temperature maintenance. When large lots of bees are incubated, the heat from bee larvae can be high enough to raise the room temperature to the point of actually killing the larvae. Therefore, adequate cooling, as well as heating, is required during the incubation process. The level of humidity is not nearly as important as maintaining the temperature level. Newly emerged bees or those still in cocoons can be held as long as a week at reduced temperatures (55° to 60°), if inclement weather occurs during the release schedule.

Protection From Parasites and Predators

Parasite-predator control during incubation should be exercised, especially in the loose cell method, as a 5-percent infestation could easily increase to 80 to 90 percent without some type of control. Most leafcutter bee parasites can complete two generations during the time required for normal emergence of the bees. Adequate control of pests during incubation has been obtained through the use of sprays, repellents, light, or emergence traps.

The leafcutter bee is attacked by numerous insects, and considerable attention must be given to maintenance of a pest-free population if an increase in bee populations is to be realized. Table 2 lists the common insects associated with leafcutter bee nests in some parts of the United States.

Probably the most important type of parasite-predator control is the maintenance of clean bee stocks by excluding pest populations through changing nesting media yearly or by utilizing emergence traps. Most pest species can be controlled during incubation or emergence through the use of sprays or traps.

Currently, chalk brood, a disease associated with bee larvae, is increasing. In some Western States, the incidence of this disease has increased to as high as 80 percent of the overwintered bee larvae. However, little is known of the causal organism and its taxonomic status. We still do not know whether the organism is the cause or merely a symptom of these bee losses. Until these questions are adequately researched, control measures cannot be devised. However, it has been shown that growers who use clean nesting media have less chalk brood than those who reuse infested nesting media.

Additional Practices

It can be highly advantageous for honey bee keepers to handle leafcutter bees in addition to their honey bee colonies. The cost of maintaining and increasing bees is much less than the potential income from selling bees or their services. New bee boards sell for about $5 each and are $50 to $70 each when covered with bees. Cells removed from bee boards or boxes (loose cells) sell at an average of $100 a gallon (10,000 cells a gal). Many individuals have made a high profit by setting out bee boards on old outbuildings, cliff banks, and other likely nesting sites and collecting the filled boards for resale in the fall. Beekeepers also might consider custom pollination—providing the pollinators and shelters for fields during the flowering season for a percentage of the crop.

Other Leafcutter Bees

Megachile concinna, the Pale Leafcutter Bee

This species is quite similar to the alfalfa leafcutter bee, *M. rotundata*. Except for minor structural differences, they are hard for the novices to distinguish. In California, these two species often are confused. The pale leafcutter bee is found in the warmer Southern and Southwestern States. It is an African species that was accidentally introduced into North America late in the 19th century. This species has not been managed like *M. rotundata*, primarily because its nesting requirements have not been adequately researched. It is capable of producing almost a complete second generation, and females have a tendency to nest in places other than the nesting shelter, such as holes in the ground. However, recent field tests indicate that *M. concinna* is an exceptionally good alfalfa pollinator. In field tests *M. concinna* made more cells and the plants produced more seed than did *M. rotundata* in similar tests.

M. apicalis and M. leachella

Both of these Eurasian species are found in the United States, but infrequently. They are potential

TABLE 2.—*Characteristics of nest associates of the alfalfa leafcutter bee*

	Native +; introduced 0	Predators +; parasites 0; scavengers —	Important +; minor 0	Control measures
Moths				
Plodia interpunctella	0	+ 0	0	Light traps.
Vitula edmandsae	+	+ 0	0	Light traps.
Wasps				
Sapyga pumila	+	0	+	Emergence traps.
Monodontomerus obscurus	0	0	+	Emergence sprays.
Monodontomerus montivaga	+	0	+	Emergence.
Pteromalus venustus	0	0	+	Emergence.
Tetrastichus megachilids	+	0	+ 0	Emergence sprays, traps.
Melittobia chalybii	+	0	0	Destroy.
Dibrachys maculipennis	0	0	0	Destroy.
Leucospis affinis	+	0	0	Nesting media.
Vespula spp	+	— +	0	Traps.
Formica spp	+	— +	0	Barrier.
Bees				
Coelioxys funeraria	+	0	0	Early emergence?
Coelioxys gilensis	+	0	0	(?)
Coelioxys	+	0	0	(?)
Stelis sp	+	0	0	(?)
Diptera flie				
Anthrax irroratus	+	0	0	Emergence.
Beetles				
Nemognatha lutea	+	+	0	Eliminate host plants.
Trichodes ornatus	+	+	+	Traps.
Ptinus californicus	+	—	0	Sprays, loose cell.
Trogoderma glabra	+	+ —	0	Cold treatment, traps.
Trogoderma variabile	+	—	0	Baits.
Tribolium castaneum	+	—	0	Sanitation.
T. audox, T. brevicornis	+	—	0	Sanitation.
Oryzaephilus surinamensis	0	—	0	Sanitation.
Cryptolestes ferrugineus	0	—		Sanitation.
Tenebroides mauritanicus	0	—	0	Sanitation.
Earwigs				
Forficula auricularia	0	— +	0	Barriers.

forage legume pollinators, and populations of both are being studied as candidate alfalfa pollinators.

The Alkali Bee

The alkali bee is brightly colored and nests in the ground in dense colonies. Each female excavates its own tunnel and cells, and there is no division of labor among the progeny. The bee was quite common in many Western States, but populations recently have declined drastically. Its usefulness as a crop pollinator was first noted in the 1940's, when large populations were observed pollinating alfalfa in Utah. By the 1950's, some growers in several Western States were improving natural nesting sites and constructing artificial bee beds. In the late 1960's and 1970's, the use of alkali bees declined in most areas due to replacement by the alfalfa leafcutter bee. Recently, parasites and diseases of the leafcutter bee have reduced its advantages, and there is increasing interest in using the alkali bee for alfalfa pollination. Certain growers in eastern Washington and the San Joaquin Valley of California have relied consistently on the alkali bee as an alfalfa pollinator, and they are convinced that the alkali bee is the best alfalfa pollinator (fig. 3).

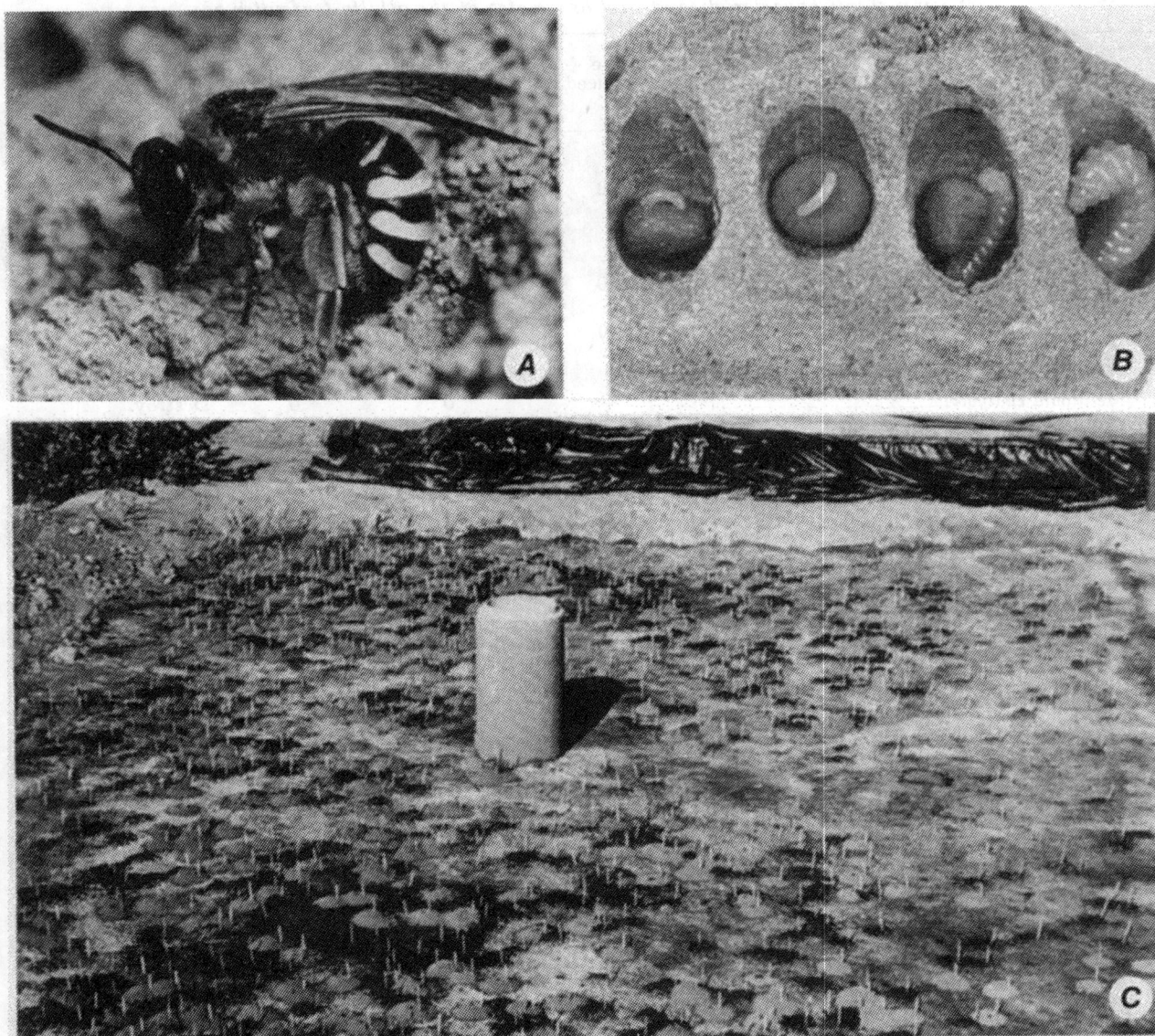

FIGURE 3.—Alkali bees: *A*, An adult female; *B*, Part of an underground brood chamber showing eggs and larvae on pollen-nectar food supply; *C*, An artificial nesting site; water is added to the bed through the drain site. Each nest (about 1,500) is marked by characteristic mound of soil.

Alkali bees collect nectar and pollen from a number of plants. The main crop plants visited are alfalfa, sweetclover, onions, and mints. Weedy plants also are visited, and these include saltcedar (*Tamarix*), morning glory (*Convolvulus*), greasewood (*Sarcobatus*), Russian thistle (*Salsola*), Rocky Mountain bee plant (*Cleome*), and several crucifers.

Alkali bees in California usually emerge in May and in Utah as late as August. Depending on location, these bees may have one or more genera-

tions per season. Males usually emerge a day or so before the females and begin patrolling the nesting area in search of newly emerged females, which they pounce upon readily. The males often are so numerous at nesting sites that they tend to discourage females from constructing nests.

On warm days, the females visit fields from about 2 hours after sunrise to 2 hours before sundown. They are capable of tripping alfalfa flowers at an average rate of 12 per minute. Normally, they provision at least one cell per day and in their lifetime provision about 12 cells (maximum 24). This foraging activity results in about 2,000 alfalfa flowers tripped per day per female and at least 25,000 flowers per female lifetime. The alkali bee visits shaded and exposed flowers, thereby increasing its pollination efficiency. The pollen and nectar provisions are stored in clusters of cells made 8 to 16 inches beneath the surface of silty loam soil.

The provisions are formed into a flattened ball with an egg deposited on the upper surface. The egg soon hatches, and successive stages of larval forms consume the provisions before entering an overwintering stage or progressing on to the pupal-adult stages and emerging to begin another generation.

Nesting Sites

Alkali bees are encouraged to nest in either natural or artificial nesting sites located in the ground (fig. 3). Generally, the soil is a silty loam, but the bees will utilize sandy soil or types with more clay particles. The most critical factor in managing alkali bee populations is the maintenance of proper soil moisture. Soil moisture levels can be measured easily by using soil tensiometers. Soil moisture in the beds and on the surface is maintained by adding salts, either calcium chloride (CaCl) or sodium chloride (NaCl), to the water. Natural nesting sites are found in seep areas where there is a constant supply of moisture extending upwards to the ground surface. Natural nesting sites can be maintained by placing a series of blind ditches to grade throughout adjacent areas and providing water for seepage to the sites.

Fencing is used to prevent packing or destruction of the soil surface by both farming operations and livestock. Weeds must be controlled to ensure a bare and attractive surface.

Artificial beds can be made readily by creating a water reservoir beneath the soil surface (fig. 3, *right*). The recipe for artificial beds calls for the following:

(1) Excavate a hole about 4 feet deep with sloping sides and construct ridges of soil in the bottom of the excavation to divide it into compartments;

(2) Line the bottom and side with thick-gage plastic sheeting and place several inches of soil on the plastic to prevent punctures;

(3) Place several cloth sacks of soil in each compartment;

(4) Add a foot of gravel on top of the soil to hold the water and install vertical pipes and perforated drain pipe in a radial pattern from the base of a stand pipe in the gravel layer;

(5) Stretch a layer of burlap on the gravel layer to prevent soil infiltration;

(6) Fill the remaining hole with soil to the surface level, forming a gentle crown to prevent ponding;

(7) Add water and salts to the stand pipe and reshape the surface as soil compaction takes place;

(8) Add brine solution to the soil surface before nesting.

Stocking of new or replenished nesting sites can be accomplished by two methods. The most common method involves digging cubic-foot blocks of undisturbed soil from a densely populated bee bed and incorporating the blocks into a new or revitalized bed before emergence of the bees in the spring. These soil blocks often contain 200 larvae each and can be purchased from growers, especially in the State of Washington. Another method of seeding an artificial bed is to sweep up newly emerged adults from natural nesting sites and release them after dark at the new site. Holes are prepared in the bed for the bees to crawl into and be protected during the night. Better results are obtained also by transplanting adults to a bed where alkali bees are active. The adults cannot be transferred to a new bed if the new area is within the old flight range (5 to 9 miles).

The area for the bee bed depends on the acreage to be pollinated. Millions of alkali bees can be produced in an area of less than 1 acre. The most ideal situation is for a community-based operation, since these bees can fly great distances and growers who build bee beds generally have no control of where their bees will forage. Alkali bees easily can fly several miles to the fields of another grower, and this long flight range makes them especially susceptible to pesticides applied in densely cultivated farming areas.

The population of alkali bees needed to set seed is highly variable and depends on the condition of the field. Most recommendations are for saturation at the level available forage can support, but it has been calculated that an acceptable level of alkali bee females per acre is between 3,500 to 5,000.

Protecting Bees from Parasites

Alkali bees are attacked by a number of insects and animals. Among the more common are the bombyliid (*Heterostylum robustum*), a sarcophagid fly (*Euphytomina nomivora*), and conopid flies (*Zondion obliquefasciatum*). Generally, high populations of these flies indicate the bed is not adequately populated with bees, since well-populated sites provide good defense by limiting oviposition by the flies. Some fly control has been obtained by traps or sprays.

A recent nest predator found in some areas is the black blister beetle (*Meloe nigra*). This beetle is easily controlled, since the flightless females must crawl from the site to deposit their eggs.

Various methods are used to discourage vertebrate predators such as birds and skunks. Federal laws now protect most animals, so growers must seek information at the local level before using control measures.

Heavy rain also can contribute to the destruction of bee cells by creating a favorable environment for soil pathogens to develop and destroy the pollen. No effective control measures have been developed to protect bees from infrequent drenching by summer rains.

Blue Orchard Bee

During the last 25 years, orchard acreages planted to stone and pome species increased dramatically. The major pollinator species for these crops is the honey bee, but—unfortunately—its colony numbers steadily decreased during the same 25-year period. We decided, therefore, to search for alternative pollinator species of these crops. A survey was initiated in 1970 and by 1972 we found a most promising species, *Osmia lignaria* Say, that was named the blue orchard bee. We have worked since to develop an effective management program for its use in orchards. Research on this species is nearing completion, and growers are rapidly becoming aware of its real potential as an alternate pollinator of orchard crops (fig. 4).

The blue orchard bee is a native species widely distributed across most of the contiguous United States and southern Canada. It has been collected in diverse habitats from sea level to elevations of 7,000 ft, but it is not found in low deserts or subtropical areas. Two allopatric subspecies are recognized, and these are isolated by the Rocky Mountains.

This robust bee flies early in the spring when few other species are active, and its size (slightly shorter than the honey bee) and color (blue-black) are diagnostic. Like the alfalfa leafcutter bee, it carries light-colored pollen on the underside of its abdomen. Females have a flat-tipped, brilliant-green prong protruding from the head directly above each mandible.

Biology

This bee nests in existing holes and fills them with a series of cells. Each female collects mud, carries it to her nest, and constructs a thin-walled partition that completely seals the cavity near its terminus. She subsequently collects nectar and pollen on each foraging trip and deposits these materials directly in front of the mud partition until the lower half of the horizontal cavity is filled with a loaf-shaped provision. An egg is deposited on the surface of the provision, and the cell is sealed by a second mud partition in front of the provision. A series of cells is thus constructed until the cavity is filled. Finally, the nest entrance is sealed with a thick plug of mud.

Eggs hatch within a week, and each larva consumes its provision for a month. A characteristic cocoon is spun, after which the larva enters a "rest" period for approximately a month. Pupation occurs in late summer, and the resultant adult remains in the cocoon through the winter. As the temperature rises in the spring, adults chew through cocoons and nest partitions, take flight, mate, and reestablish nesting.

Nesting Materials

Three nesting materials (straws, drilled holes in wood, and drilled holes in stryofoam) have been tested for trapping populations and establishing them in orchards. The blue orchard bee will nest in all three materials, but drilled wood is the most attractive for nests. The preferred hole depth is 5 to 6 inches, and the most attractive size of hole is %$_{32}$-in diameter.

Trapping

In Utah, the best areas for trapping this species are in undisturbed drainages with mixed stands of maple and aspen interrupted by open meadows. Trapping is also good on the periphery of agricultural areas in habitats that include a ready supply of mud, early bloom, and dead trees with beetle holes.

FIGURE 4.—Blue orchard bee: *A*, Adults and cocoons (pupae); *B*, collecting pollen from an apple blossom; *C*, two eggs laid on a pollen ball in the nest, cells are divided by discs of mud.

Trap nests, each having 30 to 60 holes, are attached horizontally to old stumps early in the year before nesting activities. The traps are removed 1 month following the initiation of nesting and stored until the following year in shaded, outdoor areas having good ventilation.

Nest Shelters

Experiments completed to date in commercial orchards indicate that the best size for nest shelters is a boxlike structure approximately 32 by 16 by 12 inches (wooden military foot lockers serve as ideal shelters). The shelters are attached to metal fence posts and faced in a southeast direction to catch the morning sun. Each shelter is filled with trap-nest materials. Several shelters should be used for each acre pollinated.

Population

Cage studies on almond demonstrate that fewer than 10 female *lignaria* bees will adequately pollinate a tree. While this cannot be applied directly to open pollination conditions, the information can be used to indicate the approximate number of blue orchard bees required to pollinate each acre. Since males of this species also visit blooms frequently and are proved pollinators, we estimate that 600 nesting females will adequately pollinate 1 acre of fruit trees.

Commercial Utilization

Field populations of the blue orchard bee have been tested only recently, but results of such studies justify comments on the advantages of this species as an orchard crop pollinator. These are:

(1) Nesting is completed during the short period of orchard bloom when pesticides normally are not applied. As a consequence, growers can easily schedule post-bloom pesticide schedules without affecting their pollinator force.

(2) The blue orchard bee initiates daily flight at temperatures 2° to 5° lower than for the honey bee, and individuals visit orchard bloom consistently throughout the daylight hours. This can be of much significance during "bad" weather years.

(3) Both sexes of *O. lignaria* visit orchard bloom throughout their adult lives. Therefore, the entire population serves as a pollinator force.

(4) This species increases its nesting efficiency by flying short distances to collect pollen and nectar provisions when bloom is adequate. Proper management, therefore, can be applied to better guarantee full pollinator utilization on individual properties.

(5) The normal adult flight periods of this species are short in comparison to those of other pollinators, and its annual appearance is early in the spring when inclement weather conditions are common. Nevertheless, this species is successful because females rapidly collect pollen over long daily flight periods. These facts, together, indicate the bee can serve as an effective pollinator of massive plants, such as orchard crops, that have short flower life.

(6) The blue orchard bee has its complement of nest associates, but these species have been nearly eradicated in field population tests by applying proper management techniques.

(7) The nests and nesting materials used for this species are relatively light. Therefore, management of large field populations will require fewer pieces of expensive equipment designed to move and transport heavy pollinator units.

(8) Few worker-hours, relatively, are needed for management of the species and this is restricted to a short period annually.

(9) Necessary transport of nesting populations throughout the summer months is eliminated in managing the blue orchard bee because it is an obligatory one-generation species.

(10) *Osmia lignaria* can successfully nest and pollinate crops in competition with other pollinator species visiting the same crops.

If the acreage of orchard crops continues to expand as it has in the last 25 years, it soon will be impossible to supply enough pollinators for these crops, regardless of the species used. This problem can be resolved, however, by using a multiple-species approach. Further, those presently involved in honey bee pollination could easily adapt their operations to accommodate management of additional species for concomitant use. The management of multiple species already has proved successful for several western honey bee suppliers who have added the alfalfa leafcutter bee to their operations.

Related Species

Sibling species of the blue orchard bee occur in other parts of the world. One of these, *Osmia cornifrons*, occurs in Japan and is an established pollinator of apple in that country. We cooperated with Japanese scientists recently in working with this species in Utah. While the study population nested well under Utah conditions, its progeny did not overwinter successfully. Another SEA scientist is working with this species in the Washington, D.C., area to determine if climatic conditions on our east coast (similar to Japan's) will permit successful overwintering of *O. cornifrons*.

Populations of the European species, *Osmia cornuta*, have been received from a Spanish collaborator, and we have completed studies of its biology and pollinator potential. Unlike the Japanese species, *cornuta* overwintered well under Utah conditions, and we are increasing our greenhouse-reared populations for possible field release in the near future.

Other European species are being established in our Utah greenhouses, and their status as potential pollinators has yet to be deciphered.

Other Pollinators

Bumble Bees

Bumble bees are important pollinators of many of our native plants, especially those growing at high elevations. Also, bumble bee species are associated with our crop plants, but no successful attempts have been made to utilize them on a large scale. The number of workers per colony is quite low (50–100), and the profitability of using such small colonies of bumble bees is unknown (fig. 5). Researchers, however, are studying means of increasing colony size and extending the life of colonies to more than a single season.

The generalized life cycle of a colony of bumble bees is as follows: the mated queen overwinters in some type of hybernaculum. In the spring, she becomes active and begins searching for a nesting site to make cells, lay eggs, and begin a colony. After nest establishment and the emergence of the first workers, the queen enlarges the colony with the aid of the numerous workers so that its size increases during the season. By fall, new queens, drones, and workers are being produced. Then the new queens mate, leave the colony for overwintering sites, and the old queen, workers, and drones gradually die.

Bumble bees are quite cyclic in their abundance, and this periodicity has been linked to cycles of rodent populations in whose old nests many of these bees find suitable nesting sites. Several bumble bees are quite common, and can be trapped by placing nesting boxes out in spring or by catching overwintered females before they begin to build their nests in the spring. Many of the captured females will begin to make colonies in these boxes, if pollen and nectar are provided. After the queen begins making cells and the first workers appear, the box can be opened and the bees allowed to forage in the field.

Anthophorids

Several species of Anthrophoridae have the potential of being used as crop pollinators. Although they nest in the ground, artificial nesting sites have been made from blocks of adobe, which are readily accepted by these bees. Excellent nesting and reproductions by some native species such as *Anthophora urbana*, *A. pacifica*, and *A. bomoides* have been obtained in greenhouses, but these species have not yet been field tested. A Polish species, *A. parietina*, has been successfully tested in hairy vetch in that country.

Other anthophorid bees are potential pollinators of such crops as cotton (*Diadasia* and *Ptilothrix*) and sunflowers (*Diadasia* and *Melissodes*). These bees possess characteristic habits such as gregarious nesting and restricted floral visitation, which make them potential crop pollinators. No research, however, has been done on utilizing them as such.

There are several genera of bees, such as *Peponapis* and *Xenoglossa*, whose sole host plants are species of native and cultivated squash. Some

FIGURE 5.—Bumble bees: *Left,* Four common species of western bumble bees; *right,* large nest of *Bombus morrisoni.*

species nest in dense aggregations and might be managed somewhat like our alkali bees. However, no management programs have been developed to utilize these efficient pollinators on a commercial scale.

Introduction of Foreign Pollinators

There have been only two intentional introductions of pollinators from other countries to the United States. The most obvious and successful is the honey bee. Its introduction has been so successful that additional introductions have been largely ignored. The other example is a group of fig wasps without which hybrid figs cannot be produced (except with hormonal sprays).

Several additional species of bees and wasps have become established in the United States, mostly inadvertently via our numerous vehicles of commerce. One, the alfalfa leafcutter bee, has become the most important of these adventive species. Many other species of potential crop pollinators are known to exist in foreign countries and might be valuable supplements to our existing species.

A principal shortcoming of the honey bee as a pollinator of specific crops is its wide host range. This results in a large share of each colony's population not pollinating the desired crop. Thus, the best means of obtaining specific pollinators would be the introduction of oligotrophic species that have a principal host plant.

There are many examples of foreign species that have definite affinities for certain of our important crop plants. For example, alfalfa in its native home in Eurasia is effectively pollinated by several genera and species of bees that do not occur in the United States (fig. 6). All these species are excellent choices for an introduction program to increase seed yields in areas of poor alfalfa seed production. Promising species of tomato and red clover pollinators are known to exist in Central America and South America, but no populations have been imported for preliminary trials. With the origin of a large number of our cultivated crops being other countries, an expanded program for the introduction and establishment of exotic pollinators should be expanded rapidly.

FIGURE 6.—Eight species of Iranian bees that pollinate alfalfa.

References

ALFORD, D. V.
1975. BUMBLEBEES. 352 p. Davis-Poynter Limited. London.

BOHART, G. E.
1950. THE ALKALI BEE "NOMIA MELANDERI," a native POLLINATOR OF ALFALFA. Proceedings of the 12th Alfalfa Importers Conference, Lethbridge, Alberta: 32–35.

——— 1960. INSECT POLLINATION OF FORAGE LEGUMES. Bee World 41:57–64, 85–97.

——— 1962. INTRODUCTION OF FOREIGN POLLINATORS: PROSPECTS AND PROBLEMS. First In ernational Symposium on Pollination, Aug. 1960. Swedish Seed Association, Copenhagen Publication Committee No. 7, p. 181–188.

——— 1962. HOW TO MANAGE THE ALFALFA LEAF-CUTTING BEE ("MEGACHILE ROTUNDATA") FOR ALFALFA POLLINATION. Utah Agricultural Experimental Station Circular 144, 7 p.

——— 1966. THE NEED FOR ORGANIZED INFORMATION ON CROP POLLINATION AND POLLINATORS. Proceedings of the International Pollination Conference, Cambridge Univ., London, July 8, 1964.

——— 1966. STUDIES ON POLLINATION OF DIPLOID AND TETRAPLOID RED CLOVER VARIETIES IN UTAH. Proceedings of the International Pollination Conference, Cambridge Univ., London, July 8, 1964.

——— 1970. SHOULD BEEKEEPERS KEEP WILD BEES FOR POLLINATION? American Bee Journal 110:138.

——— 1970. COMMERCIAL PRODUCTION AND MANAGEMENT OF WILD BEES—A NEW ENTOMOLOGICAL INDUSTRY. Bulletin of the Entomological Society of America 16:8–9.

——— 1971. MANAGEMENT OF HABITATS FOR WILD BEES. Proceedings of the Tall Timbers Conference, No. 3, p. 253–266, Tallahassee, Fla.

——— 1972. MANAGEMENT OF WILD BEES FOR THE POLLINATION OF CROPS. Annual Review of Entomology 17:287–312.

——— W. P. STEPHEN, and R. K. EPPLEY.
1960. THE BIOLOGY OF "HETEROSTYLUM ROBUSTUM," A PARASITE OF THE ALKALI BEE. Annals of the Entomological Society of America 53:425–435.

——— and G. F. KNOWLTON.
1964. MANAGING THE ALFALFA LEAF-CUTTING BEE FOR HIGHER ALFALFA SEED YIELDS. Utah State University Extension Service EL 104:1–7.

———G. F. KNOWLTON.
1968. ALKALI BEES—HOW TO MANAGE THEM FOR ALFALFA POLLINATION. 7 p. Utah State University Extension Service EL 78.

———W. P. NYE, and L. R. HAWTHORN.
1970. ONION POLLINATION AS AFFECTED BY DIFFERENT LEVELS OF POLLINATOR ACTIVITY. 57 p. Utah Agricultural Experiment Station Bulletin 482.

——— and T. W. KOERBER.
1972. INSECTS AND SEED PRODUCTION. Seed Biology 3:1–53.

——— D. W. DAVIS, G. D. GRIFFIN, and others.
1976. INSECTS AND NEMATODES ASSOCIATED WITH ALFALFA IN UTAH. 39 p. Utah Agricultural Experiment Station Bulletin 494.

BUTLER, G. D., JR., and M. J. WARGO.
1963. BIOLOGICAL NOTES ON MEGACHILE CONCINNA IN ARIZONA. Pan-Pacific Entomologist 39:201–206.

EVES, J. S.
1970. BIOLOGY OF MONODONTOMERUS OBSCURUS, A PARASITE OF THE ALFALFA LEAF-CUTTING BEE, "MEGACHILE ROTUNDATA." Melanderia 4:1–18.

FREE, J. B.
1970. INSECT POLLINATION OF CROPS. Academic Press, London and New York.

FRICK, K. E.
1957. BIOLOGY AND CONTROL OF TIGER BEETLES IN ALKALI BEE NESTING SITES. Journal of Economic Entomology 50:503–504.

——— 1962. ECOLOGICAL STUDIES ON THE ALKALI BEE, "NOMIA MELANDERI," AND ITS BOMBYLIID PARASITE, HETEROSTYLUM ROBUSTUM, IN WASHINGTON. Annals of the Entomological Society of America 55:5–15.

——— H. POTTER, AND H. WEAVER.
1960. DEVELOPMENT AND MAINTENANCE OF ALKALI BEE NESTING SITES. 10 p. Washington Agricultural Experiment Station Circular 366.

FRONK, W. D.
1963. INCREASING ALKALI BEES FOR POLLINATION. Wyoming Agricultural Experiment Station Circular 184.

GERBER, H. S., and E. C. KLOSTERMEYER.
1972. FACTORS AFFECTING THE SEX RATIO AND NESTING BEHAVIOR OF THE ALFALFA LEAFCUTTER BEE. 11 p. Washington Experiment Station Technical Bulletin 73.

HAWTHORN, L. R., G. E. BOHART, E. H. TOOLE, and others.
1960. CARROT SEED PRODUCTION AS AFFECTED BY INSECT POLLINATION. 18 p. Utah Agricultural Experiment Station Bulletin 422.

HIRASHIMA, Y.
1963. NOTES ON THE UTILIZATION OF "OSMIA CORNIFRONS" AS A POLLINATOR OF APPLES. Kontyu 31:280.

——— 1963. FURTHER NOTES ON THE UTILIZATION OF "OSMIA CORNIFRONS" AS A POLLINATOR OF APPLES. Kontyu 31:296.

HOBBS, G. A.
1965. IMPORTING AND MANAGING THE ALFALFA LEAFCUTTER BEE. 11 p. Canadian Department of Agriculture Publication 1209.

1967. DOMESTICATION OF ALFALFA LEAF-CUTTER BEES. 19 p. Canadian Department of Agriculture Publication 1313.

1968. CONTROLLING INSECT ENEMIES OF THE ALFALFA LEAF-CUTTER BEE, MEGACHILE ROTUNDATA. Canadian Entomology 100:871–784.

1970. ALFALFA LEAF-CUTTER BEEKEEPING, ALBERTA STYLE, 1969. Report of the Pollination Conference, 9th. Hot Springs, Ark. University of Arkansas Agricultural Extension Service, MP 127:80–83.

1973. ALFALFA LEAFCUTTER BEES FOR POLLINATING ALFALFA IN WESTERN CANADA. 30 p. Agriculture Canada Publication 1495.

HOLM, S. N.
1966. THE UTILIZATION AND MANAGEMENT OF BUMBLE BEES FOR RED CLOVER AND ALFALFA SEED PRODUCTION. Annual Review of Entomology 11:155–182.

—— and J. P. SKOU.
1972. STUDIES ON TRAPPING, NESTING AND REARING OF SOME "MEGACHILE" SPECIES (HYMENOPTERA, "MEGACHILIDAE") AND ON THEIR PARASITES IN DENMARK. Entomologica Scandinavia 3:169–180.

HOWELL, J. F.
1967. THE BIOLOGY OF "ZODION OBLIQUEFASCIATUM," A PARASITE OF THE ALKALI BEE, "NOMIA MELANDERI." 31 p. Washington Agricultural Experiment Station Technical Bulletin 51.

JOHANSEN, C. A., and J. D. EVES.
1966. PARASITES AND NEST DESTROYERS OF THE ALFALFA LEAF-CUTTING BEE. 12 p. Washington Agricultural Experiment Station Circular 469.

—— and J. D. EVES.
1969. CONTROL OF ALFALFA LEAF-CUTTER BEE ENEMIES. 10 p. Washington State University Agricultural Extension Service, EM 2631 (revised).

—— and J. D. EVES.
1970. MANAGEMENT OF ALKALI BEES FOR ALFALFA SEED PRODUCTION. Report of the 9th Pollination Conference, Hot Springs, Ark. University of Arkansas Agricultural Extension Service, MP 127:77–79.

—— JACK EVES, and CRAIG BAIRD.
1973. CONTROL OF ALFALFA LEAFCUTTING BEE ENEMIES. 10 p. Washington State University Cooperative Extension Service. EM 2631 (rev.).

—— E. C. KLOSTERMEYER, J. D. EVES, and H. S. GERBER.
1969. SUGGESTIONS FOR ALFALFA LEAF-CUTTER BEE MANAGEMENT. 8 p. Washington State University Agricultural Extension Service. EM 2775 (rev.).

KAPIL, R. P., G. S. GREWAL, SURENDRA KUMAR, and A. S. ATWAL.
1970. ROLE OF "CERATINA BINGHAMI" CKLL. IN SEED-SETTING OF "MEDICAGO SATIVA" L. Indian Journal of Entomology 32:335–341.

KLOSTERMEYER, E. C., and H. S. GERBER.
1969. NESTING BEHAVIOR OF "MEGACHILE ROTUNDATA" MONITORED WITH AN EVENT RECORDER. Annals of the Entomological Society of America 62:1321–1325.

MAETA, Y., and T. KITAMURA.
1964. STUDIES ON THE APPLE POLLINATION BY "OSMIA" I. IDEA AND PRESENT CONDITION IN UTILIZING "OSMIA" AS POLLINATORS OF APPLES IN JAPAN. Tohoku Konchu Kenkyu 1:45–52.

—— and T. KITAMURA.
1965. STUDIES ON THE APPLE POLLINATION BY "OSMIA" II. Characteristics and underlying problems in utilizing "Osmia." Kontyu 3391:17–34.

MENKE, H. F.
1964. MANAGEMENT OF ENDEMIC POPULATIONS OF ALKALI BEES FOR ALFALFA SEED PRODUCTION. p. 71–76. Annual Report of International Crop Importers Association (46th).

MICHELBACKER, A. E., P. D. HURD, and E. G. LINSLEY.
1968. THE FEASIBILITY OF INTRODUCING SQUASH BEES ("PEPONAPIS" AND "XENOGLOSSA") INTO THE OLD WORLD. 49:159–167.

MICHENER, CHARLES D.
1974. THE SOCIAL BEHAVIOR OF THE BEES. 404 p. The Belknap Press of Harvard University Press, Cambridge, Mass.

MORADESHAGHI, M. J., and G. E. BOHART.
1968. THE BIOLOGY OF "EUPHYTOMIMA NOMIIVORA," A PARASITE OF THE ALKALI BEE, "NOMIA MELANDERI." Journal of Kansas Entomological Society 41:456–473.

NYE, W. P.
1970. POLLINATION OF ONION SEED AFFECTED BY ENVIRONMENTAL STRESSES. *In* The indispensable pollinators. p. 141–144. Report of the ninth pollination conference, Hot Springs, Ark. Oct. 12–15, 1970. Arkansas Agricultural Extension Service MP 127.

—— G. D. WALLER, and N. D. WATERS.
1971. FACTORS AFFECTING POLLINATION OF ONIONS IN IDAHO DURING 1969. Journal of the Society of Horticultural Science 96 (3):330–332.

—— N. S. SHASHA'A, W. F. CAMPBELL, and A. R. HAMSON.
1973. INSECT POLLINATION AND SEED SET OF ONIONS ("ALLIUM CEPA" L.). 15 p. Utah State University Agricultural Experiment Station Research Report 6.

—— and J. L. ANDERSON.
1974. INSECT POLLINATORS FREQUENTING STRAWBERRY BLOSSOMS AND THE EFFECT OF HONEY BEES ON YIELD AND FRUIT QUALITY. Journal of the American Society of Horticultural Science 99(1):40–44.

OSGOOD, C. E.
1964. FORAGING AND NESTING BEHAVIOR OF THE LEAFCUTTER BEE ("MEGACHILE ROTUNDATA"). 104 p. Unpublished M.S. thesis, Oregon State University, Corvallis.

PACKER, J. S.
1970. THE FLIGHT AND FORAGING BEHAVIOR OF THE ALKALI BEE ("NOMIA MELANDERI") AND THE ALFALFA LEAF-CUTTER BEE ("MEGACHILE ROTUNDATA"). 119 p. Unpublished Ph. D. thesis, Utah State University, Logan.

PARKER, F. D.
1976. POTENTIALS OF EUROPEAN ALFALFA POLLINATORS. 25 p. Report of the Twenty-Fifth Alfalfa Improvement Conference, July 13–15, 1976. Ithaca, N.Y.

——— and H. W. POTTER.
1974. METHODS OF TRANSFERRING AND ESTABLISHING THE ALKALI BEE. Environmental Entomology 3(5):739–743.

——— P. F. TORCHIO, W. P. NYE, and M. PEDERSEN.
1976. UTILIZATION OF ADDITIONAL SPECIES AND POPULATIONS OF LEAFCUTTER BEES FOR ALFALFA POLLINATION. Journal of Apicultural Research 15:89–92.

PEDERSEN, M. W., and G. E. BOHART.
1950. USING BUMBLEBEES IN CAGES AS POLLINATORS FOR SMALL SEED PLOTS. Agronomy Journal 42:523.

——— G. E. BOHART, V. L. MARBLE, and E. C. KLOSTERMEYER.
1972. SEED PRODUCTION PRACTICES. Alfalfa Science and Technology Monograph 15, Chapter 32:689–720.

SCHWARZ, H. F.
1948. STINGLESS BEES ("MELIPONINAE") OF THE WESTERN HEMISPHERE. 546 p. Bulletin of the American Museum of Natural History 90.

SHASHA'A, N. S., W. F. CAMPBELL, and P. W. NYE.
1976. EFFECT OF FERTILIZER AND MOISTURE ON SEED YIELD OF ONION. Hortscience 11.425–426.

STEPHEN, W. P.
1960. ARTIFICIAL BEE BEDS FOR THE PROPAGATION OF THE ALKALI BEE, "NOMIA MELANDERI". Journal of Economic Entomology 53:1025–1030.

———
1960. STUDIES ON THE ALKALI BEE ("NOMIA MELANDERI"). II. Preliminary investigations on the effect of soluble salts on alkali bee nesting sites. Oregon Agricultural Experiment Station Technical Bulletin 52:15–26.

———
1960. STUDIES ON THE ALKALI BEE ("NOMIA MELANDERI"). III. Management and renovation of native soils for alkali bee inhabitation. Oregon Agricultural Experiment Station Technical Bulletin 52:27–39.

———
1961. ARTIFICIAL NESTING SITES FOR THE PROPAGATION OF THE LEAF-CUTTER BEE, "MEGACHILE ROTUNDATA," FOR ALFALFA POLLINATION. Journal of Economic Entomology 42:989–993.

———
1965. EFFECTS OF SOIL MOISTURE ON SURVIVAL OF PREPUPAE OF THE ALKALI BEE. Journal of Economic Entomology 58:472–474.

——— and C. E. OSGOOD.
1965. INFLUENCE OF TUNNEL SIZE AND NESTING MEDIUM ON SEX RATIOS IN THE LEAF-CUTTER BEE, "MEGACHILE ROTUNDATA." Journal of Economic Entomology 58:965–968.

SZABO, T. I.
1969. USE OF THE LEAFCUTTER BEE "MEGACHILE ROTUNDATA" FOR GREENHOUSE POLLINATION. 84 p. Unpublished thesis. University of Guelph, Ontario, Canada.

——— and M. V. SMITH.
1970. THE USE OF "MEGACHILE ROTUNDATA" FOR THE POLLINATION OF GREENHOUSE CUCUMBERS. p. 95–103. Ninth Report of the Pollination Conference. Hot Springs, Ark., University of Arkansas Agricultural Extension Service MP 127.

TELFORD, H. S., C. A. JOHANSEN, and J. D. EVES.
1972. MANAGEMENT PRACTICES AND INSECTICIDE POISONING OF "NOMIA MELANDERI" CKLL. AND "MEGACHILE ROTUNDATA" (FAB.), TWO VALUABLE POLLINATORS OF ALFALFA GROWN FOR SEED IN WASHINGTON STATE. Mededelingen Fakulteit Landbouwwetenschappen Gent. 37: 776:783.

TIRGARI, S.
1963. THE BIOLOGY AND MANAGEMENT OF THE ALFALFA LEAFCUTTER BEE, "MEGACHILE ROTUNDATA." 130 p. Unpublished thesis, Utah State University, Logan.

———
1969. EXCHANGE AND INTRODUCTION OF ALFALFA-BEE POLLINATORS BETWEEN IRAN AND THE UNITED STATES: AND THE BIOLOGY OF THESE EES. 11 p. Second National Congress of Entomology and Pest Control of Man. Sept. 1969.

TORCHIO, P. F.
1966. A SURVEY OF ALFALFA POLLINATORS AND POLLINATION IN THE SAN JOAQUIN VALLEY OF CALIFORNIA WITH EMPHASIS ON ESTABLISHMENT OF THE ALKALI BEE. 106 p. Unpublished thesis. Oregon State University, Corvallis.

———
1970. THE BIOLOGY OF "SAPYGA PUMILA" CRESSON AND ITS IMPORTANCE AS A PARASITE OF THE ALFALFA LEAFCUTTER BEE, "MEGACHILE ROTUNDATA." Ninth Report of the Pollination Conference. Hot Springs, Ark. University of Arkansas Agricultural Extension Service MP 127.

———
1972. SAPYGA PUMILA CRESSON, A PARASITE OF "MEGACHILE ROTUNDATA" (F.) (HYMENOPTERA: SAPYGIDAE: MEGACHILIDAE). I. Biology and description of immature stages. Melanderia 10:1–22.

———
1972. SAPYGA PUMILA CRESSON, A PARASITE OF "MEGACHILE ROTUNDATA" (F.) (HYMENOPTERA:

SAPYGIDAE: MEGACHILIDAE). II. Methods for control. Melanderia 10.22–30.

———
1974.   BIOLOGY AND CONTROL OF SAPYGA PUMILA, A PARASITE OF THE ALFALFA LEAFCUTTING BEE. 13 p. Utah State University Agricultural Experiment Station Research Report 16.

———
1976.   USE OF "OSMIA LIGNARIA" SAY (HYMENOPTERA: APOIDEA: MEGACHILIDAE) AS A POLLINATOR IN AN APPLE AND PRUNE ORCHARD. Journal of the Kansas Entomological Society 49(4):475–482.

——— and F. D. PARKER.
1975.   FIRST GET THE SEEDS. Utah Science 39:26–30.
WATERS, N. D.
1971.   INSECT ENEMIES OF THE ALFALFA LEAF-CUTTER BEE AND THEIR CONTROL. 4 p. Univ. of Idaho Current Information Ser. 163.
——— H. W. HOMAN, and D. W. SUTHERLAND.
1973.   RAISING ALFALFA LEAFCUTTER BEES IN IDAHO. 12 p. University of Idaho Cooperative Extension Service Bull. 538.
WILSON, E. F.
1968.   LEAF-CUTTING BEE STORAGE. 5 p. Washington State University Extension Service EM 2909.

FEDERAL AND STATE BEE LAWS AND REGULATIONS

By A. S. Michael[1]

The Federal Government has no laws or regulations on bee diseases within the United States. However, on August 31, 1922, Congress passed a law, popularly known as the Honeybee Act, restricting importing living adult honey bees into the United States. This act was amended in 1947, 1962, and 1976. The 1976 amendment, Pub. L. 94-319, includes the following:

(a) In order to prevent the introduction and spread of diseases and parasites harmful to honeybees, and the introduction of genetically undesirable germ plasm of honeybees, the importation into the United States of all honeybees is prohibited, except that honeybees may be imported into the United States—

(1) by the United States Department of Agriculture for experimental or scientific purposes, or

(2) from countries determined by the Secretary of Agriculture—

(A) to be free of diseases or parasites harmful to honeybees, and undesirable species or subspecies of honeybees; and

(B) to have in operation precautions adequate to prevent the importation of honeybees from other countries where harmful diseases or parasites, or undesirable species or subspecies, of honeybees exist.

(b) Honeybee semen may be imported into the United States only from countries determined by the Secretary of Agriculture to be free of undesirable species or subspecies of honeybees, and which have in operation precautions adequate to prevent the importation of such undesirable honeybees and their semen.

(c) Honeybees and honeybee semen imported pursuant to subsections (a) and (b) of this section shall be imported under such rules and regulations as the Secretary of Agriculture and the Secretary of the Treasury shall prescribe.

(d) Except with respect to honeybees and honeybee semen imported pursuant to subsections (a) and (b) of this section, all honeybees or honeybee semen offered for import or intercepted entering the United States shall be destroyed or immediately exported.

(e) As used in this Act, the term "honeybee" means all life stages and the germ plasm of honeybees of the genus Apis, except honeybee semen.

Any person who violates any provision of section 1 of this Act or any regulation issued under it is guilty of an offense against the United States and shall, upon conviction, be fined not more than $1,000, or imprisoned for not more than one year, or both.

The Secretary of Agriculture either independently or in cooperation with States or political subdivisions thereof, farmers' associations, and similar organizations and individuals, is authorized to carry out operations or measures in the United States to eradicate, suppress, control, and to prevent or retard the spread of undesirable species and subspecies of honeybees.

The Secretary of Agriculture is authorized to cooperate with the Governments of Canada, Mexico, Guatemala, Belize, Honduras, El Salvador, Nicaragua, Costa Rica, Panama, and Colombia, or the local authorities thereof, in carrying out necessary research, surveys, and control operations in those countries in connection with the eradication, suppression, control, and prevention or retardation of the spread of undesirable species and subspecies of honeybees, including but not limited to *Apis mellifera adansonii*, commonly known as the African or Brazilian honeybee. The measure and character of

---

[1] Deceased.

cooperation carried out under this subsection on the part of such countries, including the the expenditure or use of funds appropriated pursuant to this Act, shall be such as may be prescribed by the Secretary of Agriculture. Arrangements for the cooperation authorized by this subsection shall be made through and in consultation with the Secretary of State.

In performing the operations or measures authorized in this Act, the cooperating foreign county, State or local agency shall be responsible for the authority to carry out such operations or measures on all lands and properties within the foreign country or State, other than those owned or controlled by the Federal Government of the United States, and for such other facilities and means as in the discretion of the Secretary of Agriculture as necessary.

The first apiary inspection law in the United States was established in San Bernardino County, Calif., in 1877. By 1883, a statewide law was passed by the California legislature, and by 1906, 12 States had laws relating to foulbrood. At present, almost all States have laws regulating honey bees and beekeeping.

State laws and regulations relating to honey bees and beekeeping are designed primarily to control bee diseases. Therefore, they usually attempt to regulate movement and entry of bees, issuances of permits and certificates, apiary location control and quarantine, inspection, and methods of treating diseased colonies. These laws and regulations are summarized in tables 1, 2, 3, and 4. Tables 1 and 2 are compilations of the bee laws for intrastate regulation; tables 3 and 4 are compilations of bee laws regulating interstate movement of bees and used bee equipment in the United States.

Alaska has no intrastate or interstate laws, and Missouri has no intrastate laws (tables 1–4). Also, there is a lack of uniformity in State bee laws and regulations, but considerable agreement on specific points of law. Most States require registration of apiaries, permits for movement of bees and equipment interstate, certificates of inspection, right of entry of the inspector, movable-frame hives, quarantine of diseased apiaries, notification of the owner upon finding disease, prohibition of sale or transfer of diseased material, and use of penalties in the form of fines or jail or both. Although the destruction of American foulbrood-diseased colonies is included in most State laws, table 2 shows that most States also allow the use of drugs for control or preventive treatment of this disease.

The key figure in the enforcement of bee laws and regulations is the apiary inspector. He may have the entire State, a county, or a community under his jurisdiction. His efforts are directed toward locating American foulbrood and eliminating sources of it whenever found.

The effectiveness of bee laws and regulations is based on the compliance of the beekeeper. In addition, responsibility for disease control remains with the beekeeper, who should routinely examine colonies for disease as a regular part of his management program and do what is necessary when disease is found.

State laws regulating interstate movement of bees and used beekeeping equipment vary considerably from State to State. More uniformity in interstate laws would be highly desirable and lead to a more efficient inspection service and more effective beekeeping.

TABLE 1.—*Summary of U.S. intrastate bee inspection laws and regulations, 1977*

State	Registration of		Identification of apiary required	Inspection of		Inspector		Inspection certificate required	Permit for movement of bees and equipment required
	Apiary required	Queen apiary required		Apiary required	Honey house required	Has right of entry	Must disinfect equipment		
Alabama	X		X	X		X		X	X
Alaska									
Arizona	X		X	X		X		X	X
Arkansas	X			X		X	X	X	X
California [1]	X		X	X		X		X	

See footnotes at end of table.

TABLE 1.—*Summary of U.S. intrastate bee inspection laws and regulations, 1977*—Continued

State	Registration of		Identification of apiary required	Inspection of		Inspector		Inspection certificate required	Permit for movement of bees and equipment required
	Apiary required	Queen apiary required		Apiary required	Honey house required	Has right of entry	Must disinfect equipment		
Colorado			X			X			
Connecticut	X		X			X		X	X
Delaware	X	X		X		X	X	X	
Florida			X	X	X	X		X	X
Georgia	X	X		X		X		X	
Hawaii									
Idaho	X		X	X		X	X		X
Illinois	X	X	X	X		X		X	X
Indiana	X			X		X			
Iowa				X		X			X
Kansas	X			X		X		X	X
Kentucky	X			X		X		X	X
Louisiana				X		X		X	X
Maine	X			X		X		X	
Massachusetts				X		X		X	X
Maryland	X	X	X	X	X	X			X
Michigan	X	X		X	X	X	X		X
Minnesota	X		X	X	X	X	X	X	X
Mississippi				X		X		X	X
Missouri									
Montana	X	X	X	X		X			
North Carolina		X		X		X		X	X
North Dakota	X		X	X	X	X	X	X	X
Nebraska [1]	X	X	X	X	X	X		X	X
Nevada	X	X	X	X		X	X		X
New Hampshire				X		X			
New Jersey				X		X		X	
New Mexico	X		X	X		X		X	
New York	X			X		X			X
Ohio	X	X	X	X		X		X	X
Oklahoma	X			X	X	X	X	X	
Oregon	X		X	X	X	X		X	X
Pennsylvania [1]				X		X	X	X	
Rhode Island	X	X		X		X	X		X
South Carolina						X		X	X
South Dakota	X			X	X	X	X	X	X
Tennessee	X			X	X	X	X	X	X
Texas	X	X	X	X		X		X	X
Utah [1]	X			X	X	X	X	X	X
Virginia				X	X	X	X		
Vermont	X			X		X			
Washington	X	X	X			X	X		
Wisconsin					X	X			
West Virginia	X	X	X	X		X		X	X
Wyoming	X		X	X	X	X	X	X	

[1] Wax salvage plant permitted.

TABLE 2.—*Summary of U.S. intrastate bee disease laws and regulations, 1977*

State	Owner					American foulbrood-diseased colonies or equipment						Drug treatment		Hives must have movable frames
	Must be notified	Subject to—			Has right of appeal	Must be declared	Must be quarantined	Must not be concealed	Must not be exposed	Must not be sold or transferred	Must be destroyed	Permitted for control	Permitted for prevention	
		Tax	Fees	Penalties										
Alabama	X		X	X	X		X	X	X	X	X		X	X
Alaska														X
Arizona	X						X		X	X	X			X
Arkansas	X			X		X		X	X	X	X		X	X
California	X		X	X	X		X	X	X	X	X	X	X	X
Colorado				X							X	X	X	
Connecticut	X		X	X	X	X	X	X	X	X	X		X	X
Delaware	X			X	X	X	X	X	X	X	X		X	X
Florida	X			X		X	X		X	X	X			X
Georgia	X		X	X			X	X		X	X		X	X
Hawaii														
Idaho	X	X		X	X	X	X	X	X	X	X	X	X	X
Illinois	X			X	X	X	X		X	X	X	X	X	X
Indiana								X		X	X	X	X	X
Iowa	X	X	X	X			X			X	X			X
Kansas	X	X	X	X			X		X	X	X			X
Kentucky			X	X			X							
Louisiana				X										
Maine	X		X	X		X		X	X	X	X		X	X
Massachusetts	X			X			X	X	X	X	X	X	X	X
Maryland	X			X			X	X	X	X	X	X	X	X
Michigan	X		X	X			X	X	X	X	X			X
Minnesota	X		X	X				X						X
Mississippi							X	X						
Missouri	X			X										
Montana	X		X	X			X	X	X	X	X		X	X
North Carolina	X			X			X	X	X	X	X		X	X
North Dakota	X		X	X			X	X	X	X		X	X	X
Nebraska	X	X	X	X		X	X	X	X	X	X	X	X	X
Nevada	X	X	X	X			X	X	X	X	X	X	X	X
New Hampshire	X		X	X			X	X	X	X	X		X	X
New Jersey	X			X	X		X	X		X	X	X	X	X
New Mexico	X	X	X	X	X	X	X	X	X	X	X		X	X
New York	X		X	X	X		X				X			X
Ohio			X				X		X		X			
Oklahoma	X			X		X	X			X	X		X	X
Oregon	X		X	X	X		X	X		X	X			X

	1	2	3	4	5	6	7	8	9	10	11	12	13	14	15	16
Pennsylvania	X		X	X		X			X		X	X		X	X	X
Rhode Island	X			X		X		X	X	X	X	X		X	X	X
South Carolina			X	X		X			X	X	X	X			X	X
South Dakota	X	X		X	X	X		X	X	X	X	X		X	X	X
Tennessee	X		X	X	X	X		X	X	X	X	X		X	X	X
Texas	X			X	X	X	X	X	X	X	X	X		X		X
Utah	X	X		X		X		X	X	X	X	X		X	X	X
Virginia	X		X	X	X	X		X	X	X	X	X		X	X	X
Vermont	X			X		X			X	X	X	X		X		X
Washington	X	X		X		X	X		X	X	X	X		X	X	X
Wisconsin	X			X	X	X		X	X	X	X	X		X	X	X
West Virginia	X		X	X	X	X	X	X	X	X	X	X		X	X	X
Wyoming	X			X	X	X	X	X	X	X	X	X		X	X	X

TABLE 3.—*Summary of U.S. interstate laws and regulations on entry and certification of bees, 1977*

State	Entry				Certificate			
	Prohibited	Permit required	Must be quarantined after entry [1]	Fee	Required with application	Must accompany load	Time limit for inspection [1]	Permit in lieu of
Alabama [2]	X							
Alaska								
Arizona		X			X		90	
Arkansas		X	X	X	X			
California		X				X	X	
Colorado		X			X		60	
Connecticut						X		
Delaware		X				X		X
Florida		X			X		30	X
Georgia		X			X		30	
Hawaii [2]		X	X		X			
Idaho		X		X	X			
Illinois		X			X	X	60	X
Indiana		X				X		
Iowa						X		X
Kansas [2]		X		X	X		180	
Kentucky		X					30	
Louisiana	X							
Maine						X	60	
Massachusetts		X				X	60	
Maryland [3]						X		X
Michigan	X							
Minnesota		X		X	X		60	
Mississippi		X			X	X	60	X
Missouri								
Montana		X	90 days	X		X		
North Carolina	X	X			X		60	X
North Dakota		X		X	X	X	90	
Nebraska	X	X			X	X	60	
Nevada		X	X	X		X	60	
New Hampshire		X				X		
New Jersey						X	30	X
New Mexico						X	90	
New York						X	60	
Ohio		X			X	X		
Oklahoma						X		
Oregon			30			X	60	
Pennsylvania		X	X			X	30	
Rhode Island						X		
South Carolina						X	60	
South Dakota		X		X	X	X	60	
Tennessee		X			X	X	60	
Texas		X			X	X	60	
Utah						X	30	
Virginia		X			X	X	60	
Vermont						X		
Washington		X	30			X		
Wisconsin		X			X		30	
West Virginia					X	X		
Wyoming		X		X	X	X	X	

[1] Numbers refer to days.
[2] Special requirements for used equipment.
[3] Certificate should be sent to Maryland Department of Agriculture.

TABLE 4.—*Summary of U.S. interstate laws on hive location and bee movements, 1977*

State	Locations		Notification required upon arrival [1]	Inspection stamp required on hives	Package bees		Queen bees	
	Controlled	Advance filing required [1]			Certificate required [1]	Food restrictions	Certificate required [1]	Food restrictions
Alabama					60		60	X
Alaska								
Arizona		X	X		X		X	
Arkansas		X			X		X	
California			3			X		X
Colorado			X					
Connecticut					X		X	
Delaware					X		X	
Florida		10			X		X	
Georgia					X		X	
Hawaii		--			X		X	
Idaho	X	10			X	X	X	--
Illinois			10					
Indiana		X						
Iowa								
Kansas	X	X			X		X	X
Kentucky					X	X	X	--
Louisiana								
Maine					X		X	
Massachusetts					X		X	
Maryland								
Michigan					X		X	
Minnesota	X	30					X	
Mississippi					X	X	X	X
Missouri							X	
Montana	X	X	X		X		X	
North Carolina		X			60	X	90	X
North Dakota	X	X	X		90		X	
Nebraska	X	60	X		X	X	X	X
Nevada	X	X	X		X	X	X	X
New Hampshire					X		X	
New Jersey			X		X		X	
New Mexico		X	X		X		X	
New York					X		X	
Ohio			10		X	X	X	X
Oklahoma						X		X
Oregon			5		X		X	
Pennsylvania			X	X	X		X	
Rhode Island								
South Carolina [2]								
South Dakota	X			X	X		X	
Tennessee [2]			X		X		X	X
Texas	X	10			X		X	
Utah					X		X	
Virginia		X		X	X		X	
Vermont		1			X		X	
Washington			3		X	--	X	X
Wisconsin [2]	X		5		X		X	
West Virginia [3]			60		X		X	X
Wyoming	X			X	X	X	X	X

[1] Numbers refer to days.
[2] Special requirements for used equipment.
[3] Certificate required for honey.

FEDERAL AND STATE RESEARCH, TEACHING, AND EXTENSION

By E. C. Martin [1]

Federal Research

The U.S. Department of Agriculture has a program of research on various aspects of bee management, crop pollination, diseases and pests, protection of bees from pesticides, biology of honey bees and wild bees, honey bee breeding and genetics, and honey chemistry, quality, uses, and handling. Cooperative projects also are conducted with State apiculturists. SEA research is carried on at seven laboratories. Overall guidance, coordination, and evaluation at the national level is the responsibility of the National Program Staff scientist for Crop Pollination, Bees and Honey Research.

A description of major responsibilities of each laboratory follows:

Arizona.—2000 East Allen Road, Tucson, Ariz. 85719. Pollination of agricultural crops particularly cotton, citrus, cucurbits and onions; bee nutrition and physiology, with emphasis on pollen substitutes; bee behavior and the influence on behavior of nectar quality, flower aromas, and chemical and physical stimuli; and identification and role of micro-organisms of the honey bee.

Louisiana.—Rural Route No. 3, Box 82–B, Ben Hur Road, Baton Rouge, La. 70808. Genetics and bee breeding research. Develop a selection index indicating heritability of important characteristics of bees and the combinability of characteristics between different stocks. Test new insemination methods and equipment; techniques for long-term storage of drone sperm. Study the components of aggressive behavior in honey bees; killing or early supersedure of introduced queens; liaison with Africanized bee projects in South America; control of wax moths by nonchemical means.

Maryland.—Building 476, BARC-East, Beltsville, Md. 20705. Study bee nutrition and develop fully

suitable pollen substitutes that may be made commercially available. Test drugs, antibiotics, and chemicals useful in controlling brood and adult bee diseases, wax moth, and other pests. Develop resmethrin or other suitable material for killing bees to control disease or for other purposes.

Pennsylvania.—Eastern Regional Research Center, 600 East Mermaid Lane, Philadelphia, Pa. 19118. This laboratory does not have a permanent responsibility for honey research, but has made important contributions to our knowledge of honey chemistry, quality control, and industrial uses.

Utah.—UMC 53, Room 261, Utah State University, Logan, Utah 84321. Study the systematics, biology, behavior, distribution and usefulness of selected species of wild bees. Attempt to manage species useful in crop pollination. Test the usefulness of wild bees where pollination problems exist with specific crops.

Wisconsin.—Room 436, Russell Laboratories, University of Wisconsin, Madison, Wis. 53706. Pollination of agricultural crops, particularly soybeans, carrots, onions, and cranberries; nectar secretion in soybeans; bee behavior related to production of hybrid crops; effect of environmental electricity on bees in the hive and the field; and management of colonies for most effective pollination. Systems of management and testing bee stocks for honey production; efficient methods of wintering bees; using pollen supplements; and handling package bees.

Wyoming.—P.O. Box 3168, University Station, Laramie, Wyo. 82071. Effects of pesticides on bees and finding how to reduce serious losses caused by specific pesticide-use programs; effects of low-level (sublethal) exposure to pesticides on performance of honey bee colonies; methods of residue analysis for the presence of pesticides in bees, honey, and pollen; basic studies of toxicology and mode of action of pesticides on bees. Distribution, methods of spread, and control of chalkbrood.

---

[1] Staff scientist, Science and Education Administration, retired.

State Research, Teaching, and Extension

An important aspect of the educational system of the United States is the coordination of teaching, research, and extension in the college of agriculture of the land-grant university of each State. Under this system, the State agricultural experiment station and the cooperative extension service are integral parts of each college of agriculture. Specialists attached to subject-matter departments transmit research information from the university through the county agents of the extension service or directly to people engaged in agricultural pursuits. Much credit for America's resourcefulness in agricultural production is due to involvement of the land-grant university system with agricultural industries. In small industries like apiculture, the system does not function in States where there is no qualified apiculturist at the State university.

Of the 50 States, 14 of them have a reasonably comprehensive apiculture program at the land-grant university. In addition to this, personnel from four Federal laboratories located on university campuses teach a university course and some participate in training graduate students. In a few States, university faculty with other major responsibilities teach a university course or assume some responsibility for apicultural extension work, or both. A comprehensive apicultural program may not be warranted in every State, but several important beekeeping States have little or no apicultural work at the State university and some States with highly qualified apiculturists provide little financial support for research. Fortunately, beekeepers are well organized into national, State, and regional associations and well served by newsletters and trade journals, so that useful information is widely disseminated nationwide to those who support their associations and read the journals. A comprehensive listing of apicultural workers appears in *Gleanings in Bee Culture* each year under the caption "Who's Who in Apiculture" (Root *1977*).

In addition to teaching formal courses, apiculturists at some universities accept graduate students as assistants. From these graduate students come the next generation of research and extension specialists, and some of them enter the industry in various capacities. Training in research methods and responsibilities is an important aspect of graduate work. Much of the apicultural research accomplished in the universities is done by graduate students under the guidance of their major professors. This method is capable of great productivity, providing the major professor has money to provide assistantships for graduate students. Funding of apicultural programs in most States has been consistently low. Survival of the system is important to the future of beekeeping and crop pollination and worthy of industry support.

References

Root, John.
  1977. who's who in apiculture. Gleanings in Bee Culture 105 (4):156–158, 177.

COLLECTIONS OF BEE LITERATURE

By Julia S, Merrill and Eric H. Erickson[1]

In the United States, the best known collections of bee literature are at Technical Information Systems, the Universities of California, Minnesota, and Wisconsin, and Cornell. Canada has an extensive collection at the University of Guelph in Ontario. Smaller collections can be found at the University of Illinois, Ohio State University, and elsewhere. Scientists in apiculture, commercial beekeepers, and hobbyists often possess sizable collections. Collections described in this chapter are limited only to those that the authors consider prominent or unique.

U.S. Department of Agriculture

Technical Information Systems (TIS) maintains what is probably the largest collection of bee literature in the United States and probably the world (fig. 1). The apicultural collection began as a modest library first assembled for the exclusive use of his research staff by James I. Hambleton, chief, Division of Bee Culture, Bureau of Entomology, U.S. Department of Agriculture, in 1925. This collection became the nucleus of the Bee Culture Branch of TIS. After years of growth, including the evolution of a unique beekeeping bibliography (Merrill 1976) and several administrative changes, these materials were

consolidated with the main TIS collection in June 1976.

All items are available upon request to Technical Information Systems, Lending Division, Beltsville, Md. 20705. No loans are made to individuals other than U.S. Department of Agriculture staff. Others may borrow on interlibrary loan through their resource library. Specific articles, however, may be copied upon request to the Lending Division at a current rate of $1 for each 10 pages or fraction.

TIS has over 7,000 apicultural monographs, bound periodicals, and theses, including a number of historical value. New volumes are added regularly. Fusonie and Fusonie (1976) compiled a pre-1870 selected bibliography. At one time, 135 bee journals were received from throughout the world. Now, less than 100 titles are available, more than half of which are in foreign languages. Photocopies of translations filed in TIS may be purchased at costs previously stated. Lists of these translations have appeared in *Bee World*, a publication of the International Bee Research Association. Many of the translations were prepared by the staff of the Bee Culture Branch. Since the collection has been merged with the holdings of TIS, no further translations are being made by library employees.

University of California

Shields Library at the University of California, Davis, contains approximately 1,600 monographs and 337 bound journals pertaining to apiculture, along with numerous calendared records, photographs, and clippings. Ninety volumes are rare. The library currently receives 55 foreign and 7 American bee journals.

The rare John S. Harbison manuscript collection dating from 1857–1912 is an important part of this library. Harbison was one of the first beekeepers to import honey bees into California. This collection includes records of these shipments from New

[1] Merrill, Technical Information Systems (retired), and Erickson, Science and Education Administration, Bee Management and Entomology Research, University of Wisconsin, Madison, Wis. 53706. Erickson gratefully acknowledges the following individuals for providing current information essential for this revision: Evelyn L. Gish, entomology librarian, University of Minnesota, Minneapolis; Lawrence R. Goltz, A. I. Root Company, Medina, Ohio; Samuel F. Lewis, associate director and agricultural librarian, Steenbock Memorial Library (Agriculture), University of Wisconsin, Madison, Henry T. Murphy, librarian, Albert R. Mann Library, Cornell University, Ithaca, N.Y.; Nelson A. Piper, librarian, University of California, Davis; Gordon F. Townsend, head, Department of Apiculture, and Tim Saver, librarian, McLaughlin Library, University of Guelph, Ontario, Canada, and Howard Veatch, Dadant & Sons, Inc., Hamilton, Ill.

FIGURE 1.—The Technical Information Systems is housed in the National Agricultural Library building, Beltsville, Md.

York to Sacramento by boat, freight receipts, records of other business transactions, and patent grants. Personal collections of retired professors of apiculture and others are a part of the apiculture library.

Dadant & Sons, Inc.

The collection of bee literature at Dadant & Sons, Inc., Hamilton, Ill., was started by Charles Dadant, founder of the firm, and his son C. P. Dadant, both of whom maintained worldwide contact with beekeepers and scientists. Special interest in the library held by C. P. Dadant was continued by his son M. G. Dadant and resulted in the addition of many rare and valuable editions. M. G. Dadant was responsible for developing a facility to house the collection.

Succeeding members of the firm have continued to maintain the library that now contains over 1,400 monographs, including 55 rare books and early editions. There are 66 periodicals consisting of 542 bound volumes and 561 unbound issues,

a number of which are in complete sets. Also maintained are newsletters from 25 States, thousands of State, Federal, and foreign bulletins and reports, as well as hundreds of beekeeping supply catalogs, many of which are now historical. Finally, there is a large miscellaneous collection of letters, postcards, clippings, and photographs from L. L. Langstroth.

In 1975–76, the library was reorganized and is currently cataloged. New volumes are added when considered appropriate. From the beginning, the collection has been maintained primarily as an aid in the research programs of the company and as a reference source for Dadant publications. Items may not be borrowed from the library, and public access is restricted because of the nature of the daily business routine and space limitations.

University of Minnesota

The apiculture collection at the University of Minnesota began as the personal collection of Fr. Francis Jaeger, the first chief (1913) of the Division of Bee Culture, Department of Agriculture, on the Saint Paul campus. Upon Fr. Jaeger's retirement in 1929, the one-man division of bee culture was combined with the Department of Entomology and Economic Zoology. His collection was purchased by the department for the entomology library.

Fr. Jaeger's library was composed of approximately 600 monographs, pamphlets, bulletins, and several periodicals. There is an extensive group of books from eastern Europe; publication dates range from 1634 to 1925. The collection of bee literature presently contains over 1,050 monographs, 325 of which are considered rare and kept in locked cases. There are 87 serial titles and 12 current subscriptions. Anyone may use these materials in the library. Circulation privileges are extended to faculty, staff, students, and holders of a special-privilege card.

Current monographs and periodicals are purchased under a budget allotment. Rare items are purchased with moneys from the special "Friends of Dr. Alexander Hodson Fund." Prof. Hodson was head of the Entomology, Fisheries, and Wildlife Department from 1960 until his retirement in 1974.

Cornell University

The Everett Franklin Phillips Memorial Beekeeping Library at Cornell University, Ithaca, N.Y., began in 1924. The core is a collection of valuable and rare books acquired by E. F. Phillips (deceased) when he was in charge of apiculture research at the U.S. Department of Agriculture and while he was professor of apiculture at Cornell University. Income from an endowment fund derived from gifts from beekeepers and the Dyce honey crystallization process patent provides for purchases and maintenance of the collection. The patent was assigned to Cornell University by E. J. Dyce (deceased), who was Dr. Phillips' successor. R. A. Morse, professor of apiculture, now directs the growth of the collection.

Most of the 4,500 monographs and periodicals, along with many reprints, are shelved with the general collection. Volumes printed before 1900 and other rare items are kept in locked cases in the Albert R. Mann Library. The collection is augmented by books and manuscripts donated by many individuals and associations from the United States and abroad. The most notable of these are those of Rev. L. L. Langstroth, Moses Quinby,

Evard French, John Anderson of Scotland, and some from the collection of C. C. Miller. Included are such items as the original patent granted to L. L. Langstroth for his movable-frame hive, his diary, and letters, original manuscripts of several editors of Moses Quinby's *Mysteries of Beekeeping Explained*, and manuscripts of Dr. Phillips' own books.

A. I. Root Co.

The apiculture library at the A. I. Root Company, Medina, Ohio, was started about 1870 by Root. It consists of nearly 500 monographs, along with bound volumes of *Gleanings in Bee Culture* (two complete sets) and the *American Bee Journal* (1870 forward). There are a number of rare and historical books, including some early editions by English authors, two complete sets of ABC and XYZ of *Bee Culture*, and collections of the work of L. L. Langstroth, M. Quinby, and C. C. Miller. New items are added regularly.

No provision is made for lending the materials. The library is principally a reference collection for members of the staff of the A. I. Root Co.

University of Wisconsin

The Miller Memorial Beekeeping Library was established at the University of Wisconsin, Madison, in 1923 after the death of C. C. Miller in 1920 as a memorial to that reknowned beekeeper, teacher, investigator, and author (Lewis *1974*). A committee composed of associates and admirers solicited funds, purchased Miller's private library, and presented it to the university, together with a trust fund. The income of the trust originally was for maintenance of the collection; currently, it is used only to augment the library.

In 1929, a second extensive collection was purchased from Lt. Col. H. J. O. Walker of Budleigh Satterton, Devon, England, with funds given by Sigurd L. Odegard (deceased) of Madison. The nucleus of this collection was a group of late 19th century books owned by Alfred Neighbour, an English author and manufacturer of bee equipment. Colonel Walker worked tirelessly to expand his collection until its sale.

H. F. Wilson, professor in the Department of Economic Entomology, University of Wisconsin, was custodian of the Miller Library from its inception until his retirement in 1948.

Over the years, a number of fine personal collections have been presented to the library. Many other volumes were purchased for the library by Professor Wilson out of his own funds and from moneys received from beekeepers who annually donated a 10-pound pail of honey or its cash equivalent.

The more than 2,000 rare and historical books, journals, rare manuscripts, and documents belonging to the Miller Library are housed in the Steenbock Memorial Agricultural Library, Madison campus. Under the guidance of Professor Wilson and later C. L. Farrar (deceased), S. F. Lewis, and E. H. Erickson, the entire collection grew to over 6,000 monographs and bound volumes of journals. More than 50 periodicals are received currently, of which three-fourths are in a foreign language. Monographs acquired since 1950 are shelved in the main library and are available for circulation. Rare books are seldom circulated but may be used in the library at any time with permission of the librarian. The Miller Library does not have a card catalog as the lack of funds has prohibited such an extensive undertaking. Some bibliographies are available; see references listed at the end of this section. From time to time, exceptional items are displayed on the main floor of the library.

Professor Wilson in 1930 reported that the Miller Memorial Beekeeping Library had:

1. The most extensive collection of bee literature in the English language.

2. A more complete file of bee journals in the Dutch and Flemish language than is to be found in any other single collection.

3. A more complete collection of Belgian beekeeping literature (including Flemish and French) than is to be found elsewhere.

4. The most complete collection of beekeeping literature in the German language outside the Berlin Zoological Museum.

5. The most complete collection of beekeeping literature in the French language outside the Paris Museum.

6. A more complete collection of American beekeeping literature than is to be found in any other library.

7. Several important and rare books of which only two to five copies are known.

University of Guelph

The collection of beekeeping literature at the University of Guelph, Ontario, Canada, has grown from a modest group of books and reprints in 1940 to more than 2,000 monographs (of which 80 are pre-1800), 1,000 volumes of bound journals (of which 31 are currently received), and well over 5,000 reprints. A number of valuable books are kept in the rare bookroom, where they may be consulted but are not circulated.

The core collection was placed in the Department of Apiculture by Dr. E. J. Dyce in collaboration with Dr. E. F. Phillips of Cornell University. Morley Pettit started the collection (1914–19) followed by B. N. Gates (1919–20), Eric Millen (1921–32), and Dyce. The bulk of the collection was built up by Prof. Gordon F. Townsend, of the Department of Environmental Biology and custodian of the library until about 1968, when all books were consolidated with the central campus library.

In 1973, the extensive personal library of Burton N. Gates was acquired. Unique items are Gates' personal archives, his 10,000-card apiculture bibliography, and his extensive photograph collection and historic beekeeping equipment, the last located in the Department of Environmental Biology.

A trust fund exists for current acquisitions, and additional allocations are made available for special purchases. The library attempts to purchase all new and out-of-print books as they become available.

The reprint collection has remained with the Department of Environmental Biology, where it is managed as the American Branch of the International Bee Research Association. About 5,000 reprints are presently on microfiche. Reprints may be loaned or copies made available to American members of the International Bee Research Association. A nominal fee is charged.

Michigan State University

Michigan State University's collection of early bee books originated from the donation in 1946 of a select library of historical apicultural works collected by the author, Ray Stannard Baker. Baker collected bee books for many years and described his adventures collecting bee books in *Under My Elm*, published in 1942 under his pseudonym, David Grayson. Shortly before his death, Baker donated his bee library to Michigan State University. The Baker collection is especially strong in works on bees printed in English, although it does contain early works in other languages.

Almost all of the early English books are present. It is primarily an historical collection, and most of the books were printed before 1850. It is housed in the Special Collections Division. In addition to this, the library has a good representative collection of bee books and most useful journals up to the present time. A working collection of books, reprints, and journals also is located in the library of the Department of Entomology.

Beekeeping Bibliographies

A number of bibliographies of apiculture have been prepared over the years, some considerably more comprehensive than others. Listed below are those prepared in the United States or Canada and widely recognized, along with some little-known reference lists:

U.S. Department of Agriculture.—Not long after the inception of the apicultural library, an annotated card file of selected references from the world's literature on apiculture and related topics began to facilitate reference service and literature searches. Previous publications were included, and the index remained current until it terminated in 1972. The material included spanned many fields and was cataloged under 28 major subject headings and over 200 subclasses. Copies were supplied to field laboratories of the Division of Bee Culture from Headquarters at Beltsville, Md. These laboratories at Tucson, Ariz.; Baton Rouge, La.; Beltsville, Md.; Logan, Utah; Madison, Wis.; and Laramie, Wyo., have maintained more or less complete files of these indices. This system was discontinued following the beginning of a computerized literature search service available from the Department's Data Systems Application Division, Science and Education Administration, Beltsville, Md. 20705, to all researchers in the biological sciences.

Under the auspices of the U.S. Bioenvironmental Bee Laboratory, Beltsville Agricultural Research Center, Building 476, Beltsville, Md. 20705, the original beekeeping bibliography has been microfilmed to preserve the vast store of information. Copies of these microfilms have been deposited in Technical Information Systems (TIS), Beltsville, Md., and a set (purchased by the Madison laboratory) is on loan to the Steenbock Memorial Library (Agriculture), University of Wisconsin, Madison. Copies also have been purchased by the Baton Rouge, Laramie, and Tucson laboratories and the International Bee Research Association. Other libraries may request these sets. This set of microfilms is primarily for literature searches and historical information. In addition, the Bioenvironmental Bee Laboratory in cooperation with the Communications and Data Services Division (CDSD), has undertaken the complete computerization of the bibliography. Completion of this effort is expected in 1979.

University of Guelph.—Using *Apicultural Abstracts*, a computer printout of material issued during 1950–72 was published and is available through the Department of Apiculture at Guelph. *Apicultural Abstracts* published by the International Bee Research Association, Hill House, Gerrards Cross, Bucks, SL90NR, England, is a highly comprehensive reference source.

References

FUSONIE, A. J. and D. J. M. FUSONIE
   1976. HERITAGE OF APICULTURAL LITERATURE, A BIBLIOGRAPHY OF PRE-1870 MONOGRAPHIC IMPRINTS. *In* Associates NAL Today 1(½):26–48.

LEWIS, S. F.
   1974. HOMAGE TO DR. CHARLES C. MILLER. *In* Gleanings in Bee Culture 102(12):373, 385.

MERRILL, J. S.
   1976. A SPECIAL BEE LIBRARY AND A SPECIAL BIBLIOGRAPHY. *In* Associates NAL Today 1(½):7–9.

BEEKEEPING ORGANIZATIONS

By Joseph Moffett [1]

History of National Associations

Beekeepers have had a national association almost continuously for more than 100 years. The American Bee Association, the first national organization, started at a convention in Cleveland, Ohio, in 1860.

This association met again in March and November 1861 in Cleveland. Then, it apparently disbanded because of the Civil War.

The next national association, the North American Beekeepers' Association, was founded at a convention held in December 1870 in Indianapolis, Ind. In February 1871, a second national organization, the American Beekeepers' Association, was started in a convention held in Cincinnati, Ohio.

Both groups combined to form the North American Beekeepers' Society at a joint meeting held in Cleveland, Ohio, in December 1871. This organization has existed more or less intact since then, despite several changes in name and reorganizations. It is now called the American Beekeeping Federations, Inc.

Some names that endured for several years were North American Beekeepers' Society (1871–88), North American Beekeepers' Association (1890–96), National Beekeepers' Association (1900–1920), American Honey Producers' League (1920–21), National Federation of State Beekeepers' Association (1943–48), and American Beekeeping Federation (1949 to present). In 1953, the federation became incorporated and the name was changed to reflect this.

In 1969, at the federation convention in Portland, Oreg., a producer's group formed a second national organization called the American Honey Producers. Both organizations are still active, hold annual conventions, and represent their members' viewpoints in Washington. Since each organization sometimes emphasizes a different approach or objectives, some beekeepers are members of both associations.

Accomplishments of National Associations

The national organizations are responsible for many achievements that have benefited the bee industry. One of their most important functions has been to obtain the passage of favorable legislation and to prevent the passage of unfavorable laws. A few of these past accomplishments are:

1879: Persuaded the U.S. Post Office to allow queen bees to be sent through the mail.

1906: Helped obtain passage of the Pure Food and Drug Act to prevent the adulteration of honey.

1922: Secured passage of the law that prohibits importing adult honey bees into the United States to try to prevent the introduction of acarine disease.

1949: Obtained a Federal Honey Price Support Program.

1959: Sponsored and financed a contest annually to select a honey queen. The young woman selected as queen travels extensively to promote honey.

1970: Obtained an indemnity program to reimburse beekeepers who suffer damages to their colonies from spraying with insecticides and other pesticides.

National organizations frequently have lobbied for funds to support research at Federal laboratories.

American Beekeeping Federation

Frank R. Robinson, 13637 N.W. 39th Avenue, Gainesville, Fla., is the current secretary. The federation had 1,500 members in 1977 and 910 registered at their 1978 convention in Orlando, Fla. Some activities of the federation are:

---

[1] Research entomologist, Science and Education Administration, Oklahoma State University, Stillwater, Okla. 74074.

1. Promoting the use of honey through a very active honey queen program, which includes selecting a honey queen annually and sending her on promotional tours throughout the country.

2. Publishing an informative newsletter every 2 months.

3. Sponsoring a national honey show each year.

4. Holding an annual convention to transact business and to present an educational and informative program featuring topics of interest and importance to beekeepers.

5. Representing all parts of the bee industry in Congress. Recent efforts have been made to obtain import controls, price support, and a continuation of the indemnification program.

American Honey Producers Association

Glenn Gibson, P.O. Box 368, Minco, Okla., has been executive secretary of the American Honey Producers since its founding in 1969. The producers had approximately 200 members in 1976 and 180 registered at their 1978 convention in Tucson, Ariz. This association has been producer-oriented and has concentrated on obtaining favorable legislation from Congress. Some programs they have fought for are as follows:

1. A well-funded bee research program.

2. A smooth-working program of indemnity for losses caused by pesticides.

3. A well-funded honey research program.

4. Honey price support and loans.

5. A commemorative stamp.

6. Protection from imports.

The producers also hold an informative and educational annual convention. They have sponsored several congressional receptions to present the bee industry to Congress. Congressmen and their aides are given gifts of honey at these receptions. Slides, movies, and posters are used to explain the bee industry.

American Bee Breeders Association

This national association of queen and package bee producers was founded in 1948. Annual meetings are held in conjunction with the Southern States Beekeeping Federation convention. The ABBA has established standards of reliability, and its members can use the association's emblem in their advertising. There are 40 to 50 members, but about 350 persons attend the annual meeting.

American Honey Institute (AHI)

The institute was established in March 1928 in Indianapolis, Ind., to promote the use of honey. In 1932, the office moved to Madison, Wis. Harriett M. Grace was executive director of the institute from 1939 until she retired in 1964. During this time, AHI distributed millions of pieces of literature, including newspaper releases, feature stories, and radio and TV programs. Home economists tested many recipes and published these in three well-known booklets: "Old Favorite Honey Recipes" (1941); "New Favorite Honey Recipes" (1947); and "More Favorite Honey Recipes" (1956).

The institute and the American Beekeeping Federation jointly sponsored a honey booth at the annual convention of the American Home Economics Association for many years.

In 1965, Smith-Bucklin Associates of Chicago, Ill., assumed honey promotion for the industry. The Honey Industry Council of America and the institute combined in 1967. In 1970, Smith-Bucklin suspended their activities because of lack of funds. Honey promotion by the institute has virtually stopped.

Apiary Inspectors of America

The present organization began in 1928 under the leadership of Dr. Ralph Parker of Kansas at the annual convention of the American Honey Producers League in Sioux City, Iowa. What was probably the first national meeting of apiary inspectors was held in San Antonio, Tex., in 1906, presided over by Dr. E. F. Phillips of the Department.

This organization holds an annual convention and publishes the proceedings. Membership is limited to State and Provincial Apiarists, but a recent change in the constitution permits associate members. To date, there are 10 associate members, in addition to 42 full members.

The purposes of this organization are (1) to promote better beekeeping conditions through uniform, effective laws and methods for the suppression of bee diseases and (2) to seek mutual cooperation between apiary inspection officials of different States.

Bee Industries Association

The BIA is a national organization of bee supply and equipment manufacturers. It was formed in

1941 during World War II to obtain high priority for the materials and supplies needed by the bee industry.

Now the association meets yearly during the American Beekeeping Federation convention to discuss problems unique to supply manufacturers. Examples of such problems are the redimensioning of lumber and the changing governmental regulations on the use of chemicals.

Honey Industry Council of America (HICA)

HICA was officially established in Chicago, Ill., in May 1953. It is composed of representatives of the four most important segments of the bee industry: beekeepers, honey packers and dealers, bee supply manufacturers, and bee breeders. Therefore, the council represents all phases of the industry and formulates broad national policy, in which all segments of the industry have a voice. The council also raises money to promote honey.

Of the nine council members, four are elected by the American Beekeeping Federation, two each by the National Packers and Dealers Association and the Bee Industries Association, and one by the American Bee Breeders Association.

In 1961, the council sponsored a checkoff plan in which both the honey packers and beekeepers voluntarily deducted 1 cent per 60 pounds of honey. The proceeds were used for honey promotion. The plan was not successful, however, and was discontinued in the mid-1960's.

In 1974, the threat of isomerized corn syrup as a honey adulterant caused the HICA to establish a fund to prosecute offenders and to sponsor research to find an analytical method of detecting honey adulteration. Led by HICA President Jim Powers, the organization raised more than $100,000 for a Honey Defense Fund. The council also retains an attorney to assist with legal matters related to assuring the purity of honey.

Ladies Auxiliary, American Beekeeping Federation

The ladies auxiliary meets during the federation's annual convention. More than 150 members attended the meeting in 1977 in San Antonio, Tex. An annual fall newsletter is mailed to auxiliary members. The main emphasis of the organization is support for the honey queen program for which the auxiliary raises money by selling gifts and through donations. The auxiliary also provides voluntary help for the program. A unique feature of the annual meeting is the exchange of gifts that must pertain to the honey industry.

National Honey Packers and Dealers Association

This group was organized in 1950 and reorganized in 1953. The purposes of the organization are to represent its members, to develop better methods of processing and packaging honey, and to increase the sale of honey. The association supports research and promotion by working through the Honey Industry Council of America.

In 1977, it had 33 members. The association holds a short annual meeting during the American Beekeeping Federation Convention.

International Associations

Apimondia

This international federation of beekeepers' associations consists of more than 50 national associations from most of the major countries of the world. Included are such diverse nations as the United States, Russia, Israel, Egypt, and Zambia.

Apimondia sponsors the International Apicultural Congress every second year. Papers given at the Congress usually are published as a bound book. The 26th International Congress was held in Adelaide, Australia, in 1977.

Each quarter, Apimondia publishes Apiacta, a 50-page technical journal on apiculture, in five languages—English, French, Russian, German, and Spanish. Apiacta is published in Bucharest, Romania, and Apimondia's administrative offices are at 101, Corso Vittorio Emanuele, Rome, Italy.

International Bee Research Association (IBRA)

This research association collects and abstracts all available scientific literature on apiculture throughout the world. IBRA publishes *Bee World*, a quarterly journal for beekeepers; *Apicultural Abstracts*, a compilation of scientific literature abstracts; and the *Journal of Apicultural Research*, an outlet for original bee research papers. The association also publishes books and pamphlets about bee culture.

IBRA, founded in January 1949, was known as the Bee Research Association until 1976, when international was added to the name to better describe the organization.

With the help of Dr. Gordon Townsend and the University of Guelph, Guelph, Ontario, Canada, all apicultural abstracts since 1950 have been put on computer tape by subject and author. The editor of all journals and leader of the association for more than a quarter of a century is Dr. Eva Crane.

The offices of IBRA are at Hill House, Gerrards Cross, Bucks, 5L9 ONR, England.

Regional Organizations

Southern States Beekeepers' Federation

This federation was organized in 1928 to promote Southern beekeeping. Seventeen States are represented, and fall meetings are held annually. Approximately 350 attended the 1977 convention held in conjunction with the American Bee Breeders' Association meeting. The program deals largely with package bee and queen rearing phases of the industry, and the Federation has supported research on these subjects at some southern universities.

Honey Producers' Marketing Association

This producers' marketing organization was founded during the 1960's to aid beekeepers in the North Central States in the marketing of their honey. Howard H. Schmidt of Winner, S. Dak., is president of the association. A similar organization serves the Great Lakes area.

Eastern Apicultural Society

EAS was founded in 1955 and incorporated in 1962. Both individual beekeepers and associations can be members. Most of the State associations in the Northeastern United States and the province of Ontario are members. A journal is published every 2 months.

Most members are hobbyists, and the programs are planned for the maximum benefit of this group. Several hundred attend the annual 3-day conferences, which are held in August. The 24th (1978) annual conference was at Wooster, Ohio, and the 1979 meeting was held in Ottawa, Canada.

Western Apicultural Society

WAS was founded in 1977. It is designed to serve hobbyist and sideline beekeepers and is patterned after the Eastern Apiculture Society.

State and Local Organizations

Most, if not all, States have beekeepers' associations that sponsor annual meetings, honey promotion, legislation, and beekeeper education. The oldest association is in Michigan. It was established in 1865 and has functioned continuously since that time. By 1900, at least 14 other State associations were organized.

There also are many active local associations organized on a city, county, or area basis. Some of these hold monthly meetings and are very active, such as the Cook-DuPage Association in the Chicago, Ill., suburbs.

The California Bee Breeders, Inc., has about 50 members. It started in 1933 to promote the production of quality queens and package bees in California.

California Honey Advisory Board

The advisory board operates under the California Department of Agriculture and collects money from beekeepers to promote honey and to conduct research on problems relating to bee culture. The board promotes honey through exhibits at fairs and by distributing free recipes, obtaining publicity for honey, and advertising.

A 16-minute film, "Honey, Nature's Golden Treasure," produced by the board, is shown to interested groups. The board also published several honey recipe booklets. One, "Treasured Honey Highlights," is available for a small fee from the California Honey Advisory Board, P.O. Box 32, Whittier, Calif. 90608.

The board also provides financial assistance to the University of California at Davis for apicultural research.

Other Associations

There are several other regional or multistate associations that hold meetings either regularly or occasionally. Beekeepers from Idaho, Oregon, and Washington meet from time to time as the Northwest Beekeepers. The Tri-State Beekeepers meeting has been held at Hamilton, Ill., in the summer. Beekeepers from Arkansas, Colorado,

Iowa, Kansas, Missouri, Nebraska, and Oklahoma have held joint meetings under the title of Seven States Beekeepers. The Tidewater Beekeepers Association was formed by beekeepers in the southeast coast States. Beekeepers from Iowa, Minnesota, Nebraska, and South Dakota have met as the Four States Beekeepers.

Sioux Honey Association

This beekeepers' cooperative is the world's largest honey packer and sold more than 57 million pounds of honey during the 1976 pool year. On June 30, 1977, Sioux had 994 active members.

Sioux started in 1921 in Sioux City, Iowa, and sold about 20,000 pounds of honey the first year. Edgar G. Brown, Jr., was the first president and remained in office more than 50 years. He was succeeded by Harry Rodenberg of Wolf Point, Mont.

R. F. (Barney) Remer managed Sioux for 42 years until his retirement in 1967. Robert Steele is his successor.

The Sue Bee emblem, a Sioux Indian maiden in a buckskin dress with a pair of wings, is recognized throughout the world.

The association gradually has added plants during its growth. The general offices and largest plant are in Sioux City, Iowa; five others are in Anaheim, Calif.; Umatilla, Fla.; Waycross, Ga.; Wendall, Idaho; and Temple, Tex.

ACKNOWLEDGMENT

The information for this section was obtained from the references listed and from letters and telephone conversations with officers of many of the organizations discussed. The author thanks everyone who supplied information.

References

AMERICAN BEE JOURNAL
AMERICAN BEEKEEPING FEDERATION NEWSLETTER
AMERICAN HONEY PRODUCERS NEWSLETTER
APIACTA
BEE WORLD
EASTERN APICULTURAL SOCIETY JOURNAL
GLEANINGS IN BEE CULTURE
ANNUAL REPORTS OF SIOUX HONEY ASSOCIATION
HITCHCOCK, J.D.
   1971. NONGOVERNMENT BEEKEEPING ORGANIZATIONS IN BEEKEEPING IN THE UNITED STATES. U.S. Dept. Agri., Agriculture Handbook 335:131–133.
MILUM, V. G.
   1964. HISTORY OF OUR NATIONAL BEEKEEPING ORGANIZATIONS. 88 p. American Beekeeping Federation, Urbana, Ill.

STATISTICS ON BEES AND HONEY

By D. E. MURFIELD[1]

The Department's Economics, Statistics, and Cooperative Service provides annual statistics on bees and honey. Of the 44 State statistical offices, excluding Alaska, 43 perform list maintenance, data collection, and analysis. State lists are stratified by size groups, with special emphasis placed on apiaries with 300 or more colonies.

The present statistical program consists of two reports. The first report, released in September, provides data on the number of colonies yield per colony, and honey production from commercial beekeepers with 300 or more colonies in the 20 major honey-producing States. The second report, released in January, contains annual production statistics for all States and the United States, wholesale and retail prices received by producers, value of honey production, stocks of honey in producers' hands as of December 15, and beeswax production, value, and price.

For the September report, commercial beekeepers with 300 or more colonies are surveyed. They are considered the commercial segment of primarily wholesale producers. They accounted for about 55 percent of the Nation's 1976 honey crop. Questionnaires are mailed to all names on the commercial beekeeper list, and those not responding by mail are contacted either by telephone or in person for the information. The questionnaire asks for the number of colonies at the beginning of the honey flow and the quantity of honey taken and expected to be taken during the current production year. A third question asks for information on colonies reported in one State that might be located in another State. The purpose of this question is to identify possible duplication of reported colonies between States. Approximately 7 work days elapse between data collection and publication of the September report.

For the January report, all beekeepers (regardless of size) are surveyed. The list of honey producers surveyed is developed and maintained, using various sources such as State apiarists, county agents, ASCS county offices, and State honey associations. Survey procedures for the January report are basically the same as for the September report. Questionnaires are mailed to the beekeepers about December 1 and the estimates are published about mid-January.

The major types of estimators or indications used for the two SRS honey statistics reports are the identical and control-data expansions. These indications are computed for each stratum and weighted to a State total.

The identical expansion is a ratio (current number of colonies reported divided by the number reported the previous year by the same respondents) times the previous year estimate.

The control data expansion is a ratio (universe control data divided by control data in the sample . . . those reporting) times reported sample data. Control data represent the most recent known or estimated number of colonies for each beekeeper. Beekeepers are stratified on the basis of their control data.

The total pounds of honey, as reported by producers, are divided by the sample number of colonies to obtain an average yield per colony for each stratum. A weighted State average is computed.

Stocks of honey for sale as of December 15 are obtained by dividing the reported stocks by the reported honey produced in each strata. A weighted State percentage is computed and multiplied by the estimated honey produced.

The estimate of pounds of beeswax produced is obtained by dividing the reported pounds of beeswax by the reported honey produced. A weighted State percentage is computed and multiplied by the total pounds of honey to obtain the pounds of beeswax produced.

State statistical offices consider data on number of colonies inspected by State inspectors as check data when preparing estimates for their State.

[1] Chief, Livestock, Dairy and Poultry Branch, Estimates Division, Economics, Statistics, and Cooperative Service, U.S. Department of Agriculture, Washington, D.C. 20250

TABLE 1.—*Honey: Number of colonies, production, value, and stocks, United States, 1967–77* [1]

Year	Number of colonies	Yield per colony	Honey production	Price per pound	Total value	Stocks on hand Dec. 15 for sale
	Thousands	*Pounds*	*1,000 pounds*	*Cents*	*1,000 dollars*	*1,000 pounds*
1967	4,635	46.6	215,780	15.6	33,678	56,733
1968	4,539	42.2	191,391	16.9	32,400	41,021
1969	4,433	60.3	267,485	17.5	46,742	62,743
1970	4,290	51.7	221,842	17.4	38,550	50,575
1971	4,110	48.0	197,428	21.8	43,100	30,907
1972	4,068	52.6	214,079	30.2	64,619	29,835
1973	4,103	57.9	237,647	44.4	105,434	37,663
1974	4,195	44.1	185,079	51.0	94,372	33,748
1975	4,181	47.3	197,938	50.6	100,086	32,981
1976	4,278	46.4	199,699	49.9	99,188	34,350
1977 [2]	4,315	40.9	176,309	53.0	93,394	29,914

[1] Excludes Alaska.
[2] Preliminary.

The estimates are submitted to USDA's Crop Reporting Board in Washington, D.C., for review before publication. Revisions are made only if additional data become available from later surveys. Honey statistics are subject to revision the following year and again every 5 years when reviewed in connection with the U.S. Census of Agriculture.

The estimated price of all honey sold is a weighted price obtained by combining prices for each size of container sold, both wholesale and retail. The weights are derived from reported pounds of honey sold in each category—extracted, comb, chunk, bulk, and all honey.

The statistical program on bees and honey began over 35 years ago. This program has been adjusted to meet bee industry needs for dependable statistical data. Most program changes result from expressed needs by a broad spectrum of the bee and honey industry. Programs can be modified without additional resources, if the cost of providing the new data items is offset by discontinuing a data series that is no longer needed or has limited use. The willingness and ability of the respondent to report the new or added data also must be considered.

TABLE 2.—*Beeswax: Production and value, United States, 1967–77* [1]

Year	Production	Price per pound	Total value
	1,000 pounds	*Cents*	*1,000 dollars*
1967	4,386	58.8	2,580
1968	3,797	61.6	2,340
1969	5,171	61.1	3,162
1970	4,377	60.2	2,638
1971	3,585	61.3	2,196
1972	3,988	62.1	2,476
1973	4,231	74.4	3,147
1974	3,405	114.0	3,891
1975	3,370	102.0	3,454
1976	3,361	112.0	3,777
1977 [2]	3,067	157.0	4,830

[1] Excludes Alaska.
[2] Preliminary.

HONEY PRICE SUPPORT PROGRAM

By Harry A. Sullivan [1]

Price support programs for agricultural commodities were undertaken as a result of the depression of the 1930's. For some years, price support operations were restricted to what were called the "basic" commodities (corn, cotton, peanuts, rice, tobacco, and wheat), but gradually other commodities were added.

Factors Leading to Honey Price Support Program

Sugar rationing during World War II and the requests by the Government to increase the production of honey led to a large increase in colony numbers and a proportionate increase in honey production. With the end of sugar rationing, prices for honey dropped close to prewar levels. Due to the depressed economic situation facing them, representatives of the beekeeping industry requested assistance from Congress. In taking note of the industry's request, the House Committee on Agriculture had this to say:

> Since the close of the war, the price of honey has dropped to the point where beekeepers are finding it impossible to obtain their costs of production. It appears obvious to the committee that, if these vitally important insects are to be maintained in sufficient numbers to pollinate our crops, the beekeeping industry must have immediate assistance. Until the time comes when beekeepers can receive an adequate return from pollination services, the committee believes that a price support program for honey, as provided in this bill, is the only answer to this problem.

Honey Price Support Legislation

The Agricultural Act of 1949 requires that honey, along with several other commodities under the heading "Designated Nonbasic Agricultural Commodities," be supported at a level between 60 and 90 percent of parity. In determining the actual level of support within the prescribed

[1] Agricultural economist, Price Support and Loan Division, Agricultural Stabilization and Conservation Service.

limits, the Secretary of Agriculture is directed to consider the following factors:

(1) Supply in relation to demand.

(2) Price levels at which other commodities are being supported.

(3) Availability of funds.

(4) Perishability of honey.

(5) Importance of honey to agriculture and the national economy.

(6) Ability to dispose of stocks acquired through price support operations.

(7) The need for offsetting temporary losses of export markets.

(8) The ability and willingness of producers to keep supply in line with demand.

Parity prices are a measure of the price levels needed to give agricultural commodities a purchasing power with respect to articles that farmers buy equivalent to the purchasing power of those agricultural commodities in a base period. The formula used to determine parity prices for agricultural commodities has been outlined by Congress. The parity price calculations and determinations are made by the U.S. Department of Agriculture.

Operating Features of Price Support Program

The price of honey presently is supported through warehouse- or farm-storage loans or through purchases, or both. Loans at the applicable price support rate on warehouse- and farm-stored honey are made available to beekeepers during the crop year on any or all of the honey produced during that year. By obtaining immediate cash for his crop, the beekeeper can hold his honey and market it when he thinks the price level is satisfactory to him. However, if the market price fails to rise above the support price, he may cancel his loan by delivering honey, of a value equal to the loan value, at the end of the year unless arrangements for earlier delivery have been made.

If the beekeeper has not made use of the loan feature of the program for a part of all of his production, he can then use the purchase option. The Commodity Credit Corporation (CCC) stands ready to buy at the applicable support price any of his production he wishes to sell and which is not already obligated to CCC as loan collateral.

The beekeeper obtains loans, or payment from a CCC purchase of his honey, at his Agricultural Stabilization and Conservation Service county office. The U.S. Department of Agriculture carries out price support operations through the field offices in most of the more than 3,000 counties in the country. The offices also issue delivery instructions as to time and location for honey deliveries being made to CCC.

Other Provisions of Loan and Purchase Program

Quality and Quantity Determination

For loan purposes, the beekeeper's statement is accepted as to the quality of the honey offered as collateral. He then receives 90 percent of the value (price support rate times quantity) of the collateral. When honey is actually acquired by CCC through purchase or loan default, quantity is determined by the actual weight of honey delivered. Quality is then determined by the Agricultural Marketing Service, in accordance with U.S. standards for grades of extracted honey based on samples drawn by ASCS representatives supervising delivery. CCC bears the cost of this quality and color determination.

Color and Class Differential Structure

Honey is supported on the basis of color and class. Color and class differentials for the 1977 crop are as follows:

Class and color:	Cents per pound
Table honey:	
White and lighter	33.5
Extra light amber	32.5
Light amber	31.5
Other table honey	29.5
Nontable honey	29.5

For the 1977 crop, the national average support rate on a 60-pound and larger container basis was 32.7 cents per pound.

Fees and charges

A producer pays a nonrefundable fee for each loan disbursed. The farm-stored loan fee is $10 per loan plus $1 for each lot of honey covered by the loan. The warehouse-stored loan fee is $6 plus $1 for each warehouse receipt. A delivery charge of one cent per hundredweight is assessed on the quantity of honey delivered to CCC. The producer also pays all charges relative to insurance premiums, storage, and handling.

Early Support Program

The Department first decided that mandatory honey price support could be most widely and effectively assured by working through existing marketing machinery. Under the 1950 program, packers of honey signed contracts with the Department, under which they agreed to pay beekeepers 9 cents per pound delivered to their packing plants for all the honey acquired from them that met the requirements of the program. These requirements were especially concerned with the cleanliness of the honey, its moisture content, and flavor.

The Department, in turn, agreed to accept from the contracting packer all the honey the packer offered and to pay the support price, plus established charges for handling, storage, and any processing requested by the Department.

In the 1951 season, a similar program was operated, except that a price support differential related to the degree of acceptability of honey for table use was introduced. The differential was 1.1 cents per pound between honeys of "general national acceptability" and "limited acceptability" for table use, reflecting to a degree the difference in market value for variations in this regard.

The type of price support program in operation during those first 2 years did not give universal satisfaction. The 1952 season saw the producer (i.e., beekeeper) loan and purchase agreement type of program develop, which is now in use for honey and for most of the other agricultural commodities.

Summary of Price Support Activity

As indicated in table 1, activity under the price support program was rather modest until 1964, with small quantities placed under loan and even smaller (or none) acquired by CCC. With the dropping of the official inspection requirement as a prerequisite for obtaining a loan, activity increased rapidly. A peak of 45.7 million pounds of honey placed under loan was reached in 1969. Loan activity declined subsequent to 1969, due to declining domestic production and rising honey prices. Reduced program activity led to the discontinuation of the loan provision of the program for the 1975 and 1976 crop years. However, after repeated requests by the honey industry, the loan provision was reinstated in 1977.

TABLE 1.—*Honey: Price support activity, 1950–77*

Year	Price support rate	Support as percent of parity	Quantity placed under loan		Quantity acquired by CCC
			Amount	As percent of production	
	Cents	*Percent*	*Million pounds*	*Percent*	*Million pounds*
1950	9.0	60.0	(1)	------------	7.4
1951	10.0	60.0	(1)	------------	17.8
1952	11.4	70.0	9.3	3.4	7.0
1953	10.5	70.0	3.1	1.4	.5
1954	10.2	60.0	1.5	.7	0
1955	9.9	75.0	1.9	.8	0
1956	9.7	70.0	1.6	.7	0
1957	9.7	70.0	2.9	1.2	.1
1958	9.6	70.0	5.6	2.1	.2
1959	8.3	60.0	1.3	.5	0
1960	8.6	60.0	1.1	.4	0
1961	11.2	75.0	4.2	1.5	1.1
1962	11.2	74.0	3.4	1.2	0
1963	11.2	67.0	3.2	1.1	0
1964	11.2	65.0	9.6	3.8	2.2
1965	11.2	63.0	17.3	7.1	3.3
1966	11.4	61.3	33.9	13.7	4.1
1967	12.5	64.0	31.0	13.9	5.4
1968	12.5	66.8	24.9	12.5	.1
1969	13.0	66.7	45.7	16.2	3.5
1970	13.0	63.7	40.6	17.3	(2)
1971	14.0	66.7	22.9	11.1	0
1972	14.0	62.8	19.8	9.3	0
1973	16.1	60.2	12.1	5.1	0
1974	20.6	60.0	12.5	6.7	0
1975	25.5	60.1	(3)	------------	0
1976	29.4	60.0	(3)	------------	0
1977	32.7	60.0	(4)	(4)	(4)

[1] Direct packer purchase program.
[2] Less than 100,000 pounds.
[3] Loan provision of program discontinued.
[4] Loan provision of program reinstated—data not available.

HONEY MARKETING AIDS

By J. S. MILLER [1]

The Agricultural Marketing Service (AMS) serves as the focal point for honey marketing activities of the U.S. Department of Agriculture. This Agency frequently serves as liaison between the honey industry and the various other agencies of the Department. AMS provides several marketing aids for use of the honey industry.

Current Marketing Information

Marketing of honey is nationwide, since production occurs in all States. It is estimated that approximately half of each year's honey production is sold by the producer direct to consumers through roadside stands, by house-to-house selling, through mail-order sales, or from the producer's home. Some sales are made by the small or part-time producers who have no real means of determining a true market price for their product. Often, such sales are made without factual information on the market situation.

AMS has for many years published Honey Market News to make current marketing information available to producers and other interested persons. This unbiased monthly printed report is national in scope. The report contains factual information on supply, demand, market prices, beeswax, colony and honey plant conditions, and crop production on a State and national basis. A weekly press release also is issued on a national basis; it contains prices paid to producers and importers for bulk unprocessed honey and beeswax. Information in the weekly release also is available by telephone each Friday through a telephone recorder. Data on producer, handler, broker, and packer sales of honey used in the monthly report and weekly releases are obtained from individuals or firms in the honey industry by trained AMS market news reporters. Honey import and export data are included in the monthly report to help provide a complete picture

[1] Assistant chief, Specialty Crops Branch, Fruit and Vegetable Division, Agricultural Marketing Service.

of the supply situation. Likewise, foreign honey crop reports and prices, furnished by the Department's Foreign Agricultural Service, are included to give some insight on the world honey market.

U.S. Grade Standards

U.S. grade standards for extracted and comb honey, established by the U.S. Department of Agriculture, have been effective for many years. Grade standards provide a means of uniformly classing the quality of honey. Use of the standards is not compulsory. They are designed to provide a convenient basis for sales, for establishing quality control programs, and for determining loan values. The standards also serve as a basis for inspection and grading of honey by AMS and as a quality guide for processors.

There are four designated U.S. grades (quality levels) for extracted honey: U.S. grade A, U.S. grade B, U.S. grade C, and U.S. grade D. Factors considered in determining the grade of extracted honey are flavor, absence of defects, clarity, and moisture. Honey color is classed by permanent glass color standards or by using the Pfund honey color grader.

The grades for comb-section honey are U.S. Fancy, U.S. No. 1, U.S. No. 1 Mixed Color, U.S. No. 2, and Unclassified. Grades for shallow frame comb, wrapped cut-comb, and chunk or bulk comb honey packed in tin or glass are U.S. Fancy, U.S. No. 1, and Unclassified. Factors used in determining grades are appearance of cappings, presence of pollen grains, uniformity of honey, attachment of comb to section, absence of granulation, presence of honeydew, and weight.

Inspection and Grading Services

AMS offers inspection and grading services to the honey industry on a fee-for-service basis. Four general types of service are available:

Lot Inspection.—Inspection and grading of specific lots located in plant warehouses, commercial storage, railway cars, trucks, or any other conveyance or storage facility.

Continuous Inspection.—Inspection and grading in an AMS-approved processing plant, with one or more inspectors present at all times the plant is operating to make in-process checks on preparation, processing, and packing operations.

Pack Certification.—Similar to continuous inspection, except that the inspector need not be present continuously during all operating shifts of the plant.

Unofficial Sample Inspection.—Inspection of samples submitted by an applicant for determination of grade.

Inquiries and requests for information concerning honey market news or inspection and grading services and copies of the standards for grades of honey may be addressed to Fruit and Vegetable Division, Agricultural Marketing Service, U.S. Department of Agriculture, Washington, D.C. 20250, or to field offices of the Agency.

WORLD PRODUCTION AND TRADE IN HONEY

By Gordon E. Patty[1]

The product of the bee—honey—originates almost everywhere in the United States and in all continents of the world. World output has reached 1.3 billion pounds[2] and is rising generally. The U.S.S.R. is the world's leading producer, with the United States a close second. The U.S.S.R. is mainly self-sufficient, however, and does not enter world honey trade to a significant degree as does the United States, which is now a major importer as well as producer. Other large producers, in order of importance in 1976, were Mexico, Australia, Argentina, France, Ethiopia, the People's Republic of China (PRC), and Canada. By regions, North America is first, followed by Europe, the U.S.S.R. (Europe and Asia), Africa, Asia, South America, and Oceania.

Yields per colony vary from country to country for various reasons, such as the numbers of commercial beekeepers versus hobbyists, and weather and other conditions that can change from one year to the next. For example, in 1975, yields averaged 99 pounds per hive in Australia and 92 pounds in Canada; 73 pounds in Argentina and 70 pounds in Israel. In other selected countries, yields in 1975 were as follows (in pounds per colony): Chile, 33; France, 18; West Germany, 20; Italy, 18; Japan, 46; Mexico, 44; United States, 46; and United Kingdom, 26.

World honey trade has grown to almost 300 million pounds annually. Mexico is the world's leading exporter, with other important suppliers being Argentina, the PRC, Australia, Hungary, Spain, and Canada. Europe still accounts for nearly 70 percent of all honey imports, with about 40 percent of the total going to West Germany. The United States has become a large importer and ranked second in 1975, closely followed by Japan and the United Kingdom. In 1976, West Germany imported honey valued at $43.8 million; the average price was 39.7 cents per pound.

Mexico and Argentina export large proportions of their honey production. No data are available on the PRC's total output. Australia consumes most of its production at home, exporting the surplus. Mexico's output gradually is trending upward as colony numbers increase and the use of modern hives rises. West Germany continues to be Mexico's major market, with the United States second. Although Mexico's exports to the United States have been averaging about 10 million pounds, the amount rose to an estimated 31 million pounds in 1976.

U.S. honey output has declined gradually in recent years, while imports have increased. Total U.S. supply has remained about the same. During 1976, U.S. honey imports climbed to 66 million pounds—a new record. Mexico is the main source of U.S. imports, followed by Argentina, Canada, and Australia. Recently, U.S. imports from Brazil have increased and that country was ranked fifth in 1975 and 1976. Honey imported from Mexico traditionally has been of the darker grades, but lighter types are now coming from that source. Most honey imported from Argentina and Canada is of the lighter table grades.

The United States also exports honey, but these exports are decreasing as U.S. production declines and imports grow. In 1976, the United States exported only 4.7 million pounds of honey, with West Germany, the Netherlands, and Canada the main destinations. Much of this is specialty-type honeys, whereas in former years most U.S. honey exports were in bulk.

As world honey production has risen, so has world consumption of this natural sweet. Honey increasingly is being looked on as a health food, both in this country and abroad. Although the honey bee is a valuable pollinator of farm crops, no rent is received by most of the world's beekeepers for this necessary service, leaving the production of honey as the major source of cash income for beekeepers, now and for the foreseeable future.

---

[1] Agricultural economist, Horticultural and Tropical Products Division, Commodity Programs, Foreign Agricultural Service.

[2] 2.2 pounds equal 1 kilogram (1,000 kilograms equal 1 metric ton).

GLOSSARY

By E. C. MARTIN, *National Program Staff, Science and Education Administration, retired*

Abdomen: Segmented posterior part of bee containing heart, honey, stomach, intestines, reproductive organs, and sting.

Acarapis woodi: Scientific name of acarine mite, which infests tracheae of bees.

Acarine disease: Condition caused by *Acarapis woodi*.

Alighting board: Extended entrance of beehive on which incoming bees land.

Allele: One of a pair or series of alternative genes that can occur at a given point on a chromosome.

American foulbrood (AFB): Contagious disease of bee larvae caused by *Bacillus* larvae.

Antennae: Slender jointed feelers, which bear certain sense organs, on head of insects.

Anther: Part of plant that develops and contains pollen.

Apiarist: Beekeeper.

Apiary: Group of bee colonies kept in one location (bee yard).

Apiculture: The science and art of studying and using honey bees for man's benefit.

Apis: The genus to which the honey bee belongs.

Apis mellifera: Scientific name of the Western honey bee.

Apis cerana: Scientific name of the Eastern honey bee, the honey producer of South Asia, also called *Apis indica*.

Apis dorsata: Scientific name for the large honey bee of Asia which builds open air nests of single comb suspended from tree branches, rocky ledges, etc.

Apis florea: Scientific name for the small honey bee of Asia.

Artificial insemination: See instrumental insemination.

Autopollination: The automatic transfer of pollen from anthers to stigma within a flower as it opens.

Bacillus larvae: Bacterial organism causing American foulbrood.

Balling a queen: Clustering around unacceptable queen by worker bees to form a tight ball; usually queen dies or is killed in this way.

Bee bread: Pollen stored in cells of the comb.

Bee dance: Movement of bee on comb as means of communication; best known to indicate the direction and distance of a source of nectar or pollen.

Bee escape: Device to let bees pass in only one direction; usually inserted between honey supers and brood chambers, for removal of bees from honey supers.

Bee gum: Usually hollow log hive.

Beehive: Domicile prepared for colony of honey bees.

Bee louse: Relatively harmless insect that gets on honey bees, but larvae can damage honeycomb; scientific name is *Braula coeca*.

Bee metamorphosis: The transformation of the bee from egg to larva to pupa and finally to the adult stage.

Bee moth: See wax moth.

Bee paralysis: An adult bee disease of chronic and acute type caused by different viruses.

Bee space: A space (¼- to ⁵⁄₁₆-inch) big enough to permit free passage for a bee but too small to encourage comb building. Leaving bee space between parallel beeswax combs and between the outer comb and the hive walls is the basic principle of hive construction.

Beeswax: Wax secreted from glands on the underside of bee abdomen; molded by bees to form honeycomb.

Bee tree: A hollow tree occupied by a colony of bees.

Bee veil: See veil.

Bee venom: Poison injected by bee sting.

Bee yard: (See Apiary).

Bottom board: Floor of beehive

Brace comb: Section of comb built between and attached to other combs.

Braula coeca: See bee louse.

Brood: Immature or developing stages of bees; includes eggs, larvae (unsealed brood), and pupae (sealed brood).

Brood chamber: The area of the hive where the brood is reared; usually the lowermost hive bodies.

Brood comb: Wax comb from brood chamber of hive containing brood.

Brood nest: Area of hive where bees are densely clustered and brood is reared.

Burr comb: Comb built out of place, between movable frames or between the hive bodies.

Capped brood: Brood (either last larval stage or pupal stage) that has been capped over in its cell.

Capped honey: Cells full of honey, closed or capped with beeswax.

Cappings: Beeswax covering of cells of honey which are removed before extracting.

Castes: The three types of individual bees (workers, drones, and queen) that comprise the adult population of a bee colony.

Carniolan bees: A race of honey bees which originated in the southern part of the Austrian Alps and northern Yugoslavia.

Caucasian bees: A race of honey bees native to the high valleys of the Central Caucasus.

Cell: The six-sided compartment of a honeycomb, used to raise brood or to store honey and pollen. Worker cells approximate five to the linear inch, drone cells are larger averaging about four to the linear inch.

Cell cup: Initially constructed base of queen cell; also made artificially for queen rearing.

Chilled brood: Brood that has died because of chilling.

Chromosomes: The structures in a cell that carry the genes.

Chunk honey: A jar of honey containing both liquid (extracted) honey and a piece of comb with honey.

Cleansing flight: Flight bees take after days of confinement, during which they void their feces.

Clipped queen: Queen whose wing (or wings) has been clipped for identification purposes

Cluster: Collection of bees in colony gathered into limited area.

Colony: Social community of several thousand worker bees, usually containing one queen, with or without drones. (See social insects.)

Comb: (See honeycomb).

Comb foundation: Thin sheet of beeswax impressed by mill to form bases of cells, some foundation also is made of plastic and metal.

Comb honey: Honey marketed and eaten in the comb.

Corbicula: See pollen basket.

Creamed honey: Honey made to crystallize smoothly by seeding with 10 percent crystallized honey and storing at about 57°F.

Cross pollination: Transfer of pollen between plants which are not of identical genetic material.

Crystallized honey: Honey hardened by formation of dextrose-hydrate crystals. Can be reliquefied by gentle heat.

Cut comb honey: Comb honey cut into appropriate sizes and packed in plastic.

Demaree: Method of swarm control, by which queen is separated from most of brood; devised by man of that name.

Dextrose: Also known as glucose; one of principal sugars of honey.

Diastase: Enzyme that aids in converting starch to sugar.

Diploid: An organism or cell with two sets of chromosomes, for example, worker and queen honey bees.

Disappearing disease: A condition in which colonies become weak from causes which are not readily identifiable.

Division board: Flat board used to separate two colonies or colony into two parts.

Division board feeder: A wooden or plastic trough which is placed in the hive in a frame space to feed the colony honey or sugar syrup.

Drawn comb: Comb having the cells built out (drawn) by honey bees from a sheet of foundation. Cells are about ½-inch deep.

Drone comb: Comb with about four cells to the inch and in which drones are reared.

Drone layer: A queen which lays only unfertilized eggs which always develop into drones. Results from improperly or nonmated queen or an older queen who has run out of sperm.

Dwindling: Rapid or unusual depletion of hive population, usually in the spring.

Dysentery: The discharge of fecal matter by adult bees within the *hive*. Commonly contributing conditions are nosema disease, excess moisture in the hive, starvation conditions, and low-quality food.

Escape board: Board with one or more bee escapes on it to permit bees to pass one way.

European foulbrood: Brood disease of bees caused by *Streptococcus pluton* and possibly associated organisms.

Extracted honey: Honey removed from the comb by centrifugal motion (in a special machine called an extractor) and marketed in the liquid form.

Extractor: Machine that rotates honeycombs at sufficient speed to remove honey from them.

Field bees: Those bees in the hive who are mature enough to fly from the hive on foraging missions; also termed forager bees.

Food chamber: Hive body containing honey provided particularly for overwintering bees.

Foundation: (See Comb foundation).

Frame: Rectangular, wooden honeycomb supports, suspended by top bars within hive bodies.

Fructose: (See Levulose).

Full sisters: Queen or worker bees produced by a single queen and sired by different drones that are related to each other as brothers (used in bee breeding).

Fumagillin: Antibiotic given bees to control nosema disease.

Galleria mellonella: Scientific name of greater wax moth, whose larvae destroy honeycomb.

Gamete: A male or a female reproductive cell (egg or sperm).

Gene: A unit of inheritance located at a specific location in a chromosome.

Gene pool: The genetic base available to bee breeders for stock improvement.

Germplasm: All the hereditary material that can potentially contribute to the production of new individuals.

Giant bee: (See *Apis dorsata*).

Glucose: (See Dextrose).

Grafting: The transfer of young larvae from worker cells to queen cups.

Granulated honey: (See crystallized honey).

Half sisters: Queen or worker bees produced by a single queen and sired by drones that are not related to each other.

Haploid: An organism or cell with one set of chromosomes; for example. drone bee.

Hemizygous: The condition in which only one allele of a pair is present. Drones are hemizygous at all loci.

Heterosis: Hybrid vigor.

Heterozygous: An organism with unlike members of any given pair or series of alleles (bee genetics).

Hive: Man-constructed home for bees.

Hive tool: Metal tool for prying supers or frames apart.

Hoffman frame: Self-spacing wooden frame of type customarily used in Langstroth hives.

Homozygous: An organism with identical members of any given pair or series of alleles.

Honey: Sweet, viscous fluid elaborated by bees from nectar obtained from plant nectaries, chiefly floral.

Honey bee: Genus *Apis*, family Apidae, order Hymenoptera.

Honeycomb: Comb built by honey bees with hexagonal back-to-back cells on median midrib.

Honeydew: Sweet secretion from aphids and scale insects.

Honey extractor: (See Extractor).

Honey flow: Period when bees are collecting nectar from plants in plentiful amounts.

Honey house: Building in which honey is extracted and handled.

Honey pump: Pump for transferring liquid honey, usually from the extractor to storage tanks.

Honey stomach: (Honey sac) An enlargement of the posterior end of the oesophagus in the bee abdomen. It is the sac in which the bee carries nectar from flower to hive.

Honey sump: Temporary honey-holding area with baffles usually placed between the extractor and the honey pump; tends to hold back sizable pieces of wax and comb.

Hybrid: Offspring from two unrelated (usually inbred) lines.

Hymenoptera: Order to which all bees belong, as well as ants, wasps, and certain parasitic insects.

Inbred: A homozygous organism usually produced by inbreeding.

Inbreeding: Matings among related individuals.

Inner cover: A cover used under the standard telescoping cover on a bee hive.

Instrumental insemination: The act of depositing semen into the oviducts of a queen by the use of a manmade instrument.

Introducing cage: Small wood and wire cage used to ship queens and also sometimes to release them into the colony.

Invertase: Enzyme produced by bees that speeds inversion of sucrose to glucose and fructose.

Italian bees: A race or variety of honey bee which originated in Italy and has become widely dispersed and cross-bred with other races.

Jumbo hive: Hive 2½ inches deeper than standard Langstroth hive.

Langstroth: A minister from Pennsylvania who patented the first hive incorporating bee space thus providing for removable frames. The modern hive frequently is termed the Langstroth hive and is a simplified version of similar dimensions as patented by Langstroth.

Langstroth frame: 9⅛- by 17⅝-inch standard U.S. frame.

Larva: Stage in life of bee between egg and pupa; "grub" stage.

Laying worker: Worker bees which lay non-fertilized eggs producing only drones. They occur in hopelessly queenless colonies.

Levulose: Noncrystallizing sugar of honey which darkens readily if honey is overheated.

Line breeding: Mating of selected members of successive generations among themselves in an effort to maintain or fix desirable characteristics.

Locus: A fixed position on a chromosome occupied by a given gene or one of its alleles.

Mandibles: Jaws of insects.

Mating flight: The flight of a virgin queen during which time she mates with one or more drones high in the air away from the apiary. Queens usually mate with 6 to 10 drones on two or more mating flights.

Mead: A wine made with honey. If spices or herbs are added, the wine usually is termed metheglin.

Metamorphosis: Changes of insect from egg to adult.

Migratory beekeeping: Movement of apiaries from one area to another to take advantage of honey flows from different crops.

Mite: See *Acarapis woodi* and *Varroa jacobsoni*.

Mutation: A term used to describe both a sudden change in the alleles or chromosomes of an organism and the changed form itself as it persists.

Nectar: A sweet secretion of flowers of various plants, some of which secrete enough to provide excess for the bees to store as honey.

Nectaries: Special cells on plants from which nectar exudes.

Nosema disease: Disease of bees caused by protozoan spore-forming parasite, *Nosema apis*.

Nucleus (Nuke): A small colony of bees resulting from a colony division. Also, a queen-mating hive used by queen breeders.

Nurse bees: Three-to 10-day-old adult bees that feed the larvae and perform other tasks in the hive.

Observation hive: Hive with glass sides so bees can be observed.

Ocellus (ocelli): Simple eye(s) of bees.

Package bees: A quantity of bees (2 to 5 lb) with or without a queen shipped in a wire and wood cage to start or boost colonies.

Paralysis: (See bee paralysis).

Parthenogenesis: Production of offspring from a virgin female.

Pheromones: Chemicals secreted by animals to convey information or to affect behavior of other animals of the same species. (See queen substance.)

Pistil: The combined stigma, style, and ovary of a flower.

Play flight: Short orientation flight taken by young bees, usually by large numbers at one time and during warm part of day.

Pollen: Male reproductive cells of flowers collected and used by bees as food for rearing their young. It is the protein part of the diet. Frequently called bee bread when stored in cells in the colony.

Pollen basket: Area on hindleg of bee adapted for carrying pellets of pollen.

Pollen cake: Cake of sugar, water, and pollen or pollen substitute, for bee feed.

Pollen substitute: Mixture of water, sugar, and other material, such as soy flour, brewer's yeast, etc., used for bee feed.

Pollen supplement: Pollen substitute added to natural pollen in a pollen cake.

Pollen trap: Device which forces bees entering hive to walk through a 5-mesh screen, removing pollen pellets from their legs into a collecting tray.

Pollination: The transfer of pollen from the anthers of a flower to the stigma of that or another flower.

Pollinator: The agent which transfers pollen; e.g., a bee.

Pollinizer: The plant source of pollen used for pollination; e.g., pollinizer varieties of apples and pears must be planted in order to produce a crop. Bees must carry the pollen from one variety to another.

Proboscis: Mouth parts of bee for sucking up nectar, honey, or water.

Propolis: A glue or resin collected from trees or other plants by bees; used to close holes and cover surfaces in the hive. Also called bee glue.

Pupa: Stage in life of developing bee after larva and before maturity.

Queen: Sexually developed female bee. The mother of all bees in the colony.

Queen cell: Cell in which queen develops.

Queen cup: The beginnings of a queen cell in which the queen may lay a fertile egg to start the rearing of another queen.

Queen excluder: Device usually made of wood and wire, with opening 0.163 inch, to permit worker bees to pass through but excludes queens and drones. Used to restrict the queen to certain parts of the hive.

Queen substance: Pheromone material secreted from glands in the queen bee and transmitted throughout the colony by workers. It makes the workers aware of the presence of a queen.

Race: Populations of bees, originally geographically isolated and somewhat adapted to specific regional conditions.

Ripening: Process whereby bees evaporate moisture from nectar and convert its sucrose to dextrose (glucose) and levulose (fructose), thus changing nectar into honey.

Rendering wax: Melting old combs and wax cappings and removing refuse to partially refine the beeswax. May be put through a wax press as part of the process.

Requeen: To replace a queen in a hive. Usually to replace an old queen with a young one.

Robbing: Bees steal honey from other hives. A common problem when nectar is not available in the field.

Royal jelly: Glandular secretion of young worker bees used to feed the queen and young brood.

Sacbrood: A fairly common virus disease of larvae, usually nonfatal to the colony.

Scout bees: Worker bees searching for nectar or other needs, including suitable location for a swarm to nest.

Sealed brood: Brood in pupal stage with cells sealed.

Self-pollination: The transfer of pollen from the anther to the stigma of the same flower or to flowers of the same plant or other plants of identical genetic material such as apple varieties, clones of wild blueberries, etc. (See autopollination).

Septicemia: Usually minor disease of adult bees caused by *Pseudomonas apiseptica*.

Skep: A beehive, usually of straw and dome-shaped, that lacks movable frames.

Slumgum: A dark residue, consisting of brood cocoons and pollen, which is left after wax is rendered by the beekeeper.

Smoker: Device used to blow smoke on bees to reduce stinging.

Social insects: Insects which live in a family society, with parents and offspring sharing a common dwelling place and exhibiting some degree of mutual cooperation; e.g., honey bees, ants, termites.

Solar wax melter: Glass-covered box in which wax combs are melted by sun's rays and wax is recovered in cake form.

Spermatheca: Small saclike organ in queen in which sperms are stored.

Spermatoza: Male reproductive cells.

Spiracles: External openings of tracheae through which bees breathe.

Stamen: Male part of flower on which pollen-producing anthers are borne.

Sting: Modified ovipositor of female Hymenoptera developed into organ of defense.

Sucrose: Cane sugar; main solid ingredient of nectar before inversion into other sugars.

Super: Any hive body placed above the brood chamber for the storing of surplus honey.

Supersedure: The replacement of a weak or old queen in a colony by a daughter queen—a natural occurrence.

Supersisters: Queens or worker bees produced by a single queen and sired by identical sperm from a single drone (subfamily).

Surplus honey: A term generally used to indicate an excess amount of honey above that amount needed by the bees to survive the winter. This surplus is usually removed by the beekeeper.

Swarm: Natural division of colony of bees.

Tarsus: Fifth segment of bee leg.

Thorax: Middle part of bee.

Tracheae: Breathing tubes of insects.

Tumuli: Nest mounds (wild bees).

Uncapping knife: Knife used to remove honey cell caps so honey can be extracted.

Unite: Combine one colony with another.

Unsealed brood: Brood in egg and larval stages only.

Virgin queen: Unmated queen.

Wax glands: Glands on underside of bee abdomen from which wax is secreted after bee has been gorged with food.

Wax moth: Lepidopterous insect whose larvae destroy wax combs.

Wild bees: Any insects that provision their nests with pollen, but do not store surplus edible honey.

Winter cluster: Closely packed colony of bees in winter.

Wired foundation: Foundation with strengthening wires embedded in it.

Wired frames: Frames with wires holding sheets of foundation in place.

Worker bee: Sexually undeveloped female bee.

Worker comb: Honeycomb with about 25 cells per square inch.

Worker egg: Fertilized bee egg.